Thermal and Hydro Prime Movers: Engines and Turbines

Thermal and Hydro Prime Movers: Engines and Turbines

Brent Cobb

WILLFORD PRESS
www.willfordpress.com

Published by Willford Press,
118-35 Queens Blvd., Suite 400,
Forest Hills, NY 11375, USA

ISBN: 978-1-64728-506-7

Cataloging-in-Publication Data

Thermal and hydro prime movers : engines and turbines / Brent Cobb.
    p. cm.
Includes bibliographical references and index.
ISBN 978-1-64728-506-7
1. Engines. 2. Turbines. 3. Motors. 4. Power (Mechanics) I. Cobb, Brent.
TJ250 .T44 2023
621.4--dc23

For information on all Willford Press publications
visit our website at www.willfordpress.com

WILLFORD PRESS

Contents

Preface

A prime mover refers to a type of engine that converts fuel to useful work. Inbound and outbound engines, gas turbines, steam turbines and water turbines are some examples of prime movers. Thermal prime mover and non-thermal prime mover are the two main types of prime movers. Thermal prime movers are the types of prime movers which primarily use the thermal energy from a source to produce power. Some examples of thermal primary movers are heat engines, geothermal, solar energy, nuclear power plant and bio gas. Non-thermal prime movers do not use heat energy and can be subdivided into include tidal turbines, hydraulic turbines and wind turbines. This book presents the various engines and turbines which fall under the umbrella of thermal and hydro prime movers in the most comprehensible and easy to understand language. It is appropriate for students seeking detailed information in mechanical engineering field as well as for experts.

The researches compiled throughout the book are authentic and of high quality, combining several disciplines and from very diverse regions from around the world. Drawing on the contributions of many researchers from diverse countries, the book's objective is to provide the readers with the latest achievements in the area of research. This book will surely be a source of knowledge to all interested and researching the field.

In the end, I would like to express my deep sense of gratitude to all the authors for meeting the set deadlines in completing and submitting their research chapters. I would also like to thank the publisher for the support offered to us throughout the course of the book. Finally, I extend my sincere thanks to my family for being a constant source of inspiration and encouragement.

Brent Cobb

IC Engines

1.1   IC Engines: Classification

1. Based on application:

- Aircraft Engine.

- Marine Engine.

- Automobile Engine.

- Locomotive Engine.

- Stationary Engine.

2. Based on basic engine design:

- Rotatory: Single motor, Multi motor.

- Reciprocating: Single cylinder, V, Multi-cylinder In-line, radial, opposed cylinder, Opposed Piston.

3. Based on operating cycle:

- Otto (For the Convectional SI Engine).

- Diesel (For the Ideal Diesel Engine).

- Atkinson (For complete expansion SI Engine).

- Miller (For Early/Late Inlet valve closing type SI Engine).

- Dual (For the Actual Diesel Engine).

4. Based on working cycle:

- Scavenging; direct/crankcase/cross flow; back flow/loop; Uni flow.

- Two stroke cycle.

- Naturally aspirated or turbocharged.
- Four stroke cycle.

5. Based on design of valve/port:

- Rotatory valve.
- Poppet valve.

6. Based on location of valve/port:

- L-head.
- T-head.
- F-head.
- L-head.

7. Based on Fuel:

- Crude oil derivatives; Petrol, diesel.
- Petroleum derived: CNG, LPG.
- Alternative:
  - Other sources; coal, bio-mass, tar stands, shale.
  - Bio-mass derived: alcohols, vegetable oils, producer gas, biogas and hydrogen.
- Convectional:
  - Bio-fuel and dual-fuel.
  - Blending.

8. Based on mixture preparation:

- Fuel injection.
- Carburetion.

9. Based on ignition:

- Compression Ignition.
- Spark ignition.

10. Based on stratification of charge:

- With fuel injection: Stratified charge.

- With carburetion: Homogeneous Charge.

11. Based on combustion chamber design:

- Divided chamber.

- Open chamber: Disc, wedge, hemispherical, bowl-in-piston, bath tub.

- (For CI): 1. Swirl chamber, 2. Pre-chamber.

- (for SI): 1. CVCC, 2. Other designs.

12. Based on cooling system:

- Air-cooling system.

- Water-cooling system.

	SI Engine	Cl Engine
1	Basic Cycle: SI engine is operated by Otto cycle or constant volume cycle.	CI engine is operated by Diesel cycle or constant pressure cycle.
2	Introduction of Fuel: In SI engine, the fuel is introduced to the cylinder along with air through the Inlet valve-during the suction stroke.	In Cl engine, the fuel is injected by the injector at the end of compression stroke.
3	Ignition: In SI engines, the fuel-air mixture is ignited by a high-tension spark plug. Hence it is called as spark ignition engines.	In CI engines, the ignition of fuel air mixture takes place due to the high pressure and temperature of the air. Hence, they are known compression as Ignition Engines.
4	Compression Ratio: Compression ratio for SI engine varies from 6 to 8.	Compression ratio for CI engine varies from 12 to 18.
5	Speed: These are used for high speed applications.	These are used for low speed operations.

1.1.1. Working Principles

- Cylinder Head: It carries inlet and exhaust valve. Fresh charge is admitted through inlet valve and burnt gases are exhausted from exhaust valve. In case of petrol engine, a spark plug and in case of diesel engine, an injector is also mounted on cylinder head.

- Cylinder Block: In the bore of cylinder the air-fuel mixture is freshly charged and is ignited, compressed by piston and expanded to give power to piston.

- Piston Rings: It prevents the compressed charge of fuel-air mixture from leaking to the other side of the piston. Oil rings, is used for removing lubricating oil from the cylinder after lubrication. This ring prevents the excess oil to mix with charge.

- Gudgeon Pin: Connects the piston with small end of connecting rod.

- Piston: During suction stroke, it sucks the fresh charge of air-fuel mixture through inlet valve and compresses during the compression stroke inside the cylinder. This way piston receives power from the expanding gases after ignition in cylinder. Also forces the burnt exhaust gases out of the cylinder through exhaust valve.

- Connecting Rod: It changes the reciprocating motion of piston into rotary motion at crankshaft. This way connecting rod transmits the power produced at piston to crankshaft.

- Crank Shaft: Receives oscillating motion from connecting rod and gives a rotary motion to the main shaft. It also drives the camshaft which actuate the valves of the engine.

- Inlet Valve and Exhaust Valve: Inlet valve allow the fresh charge of air-fuel mixture to enter the cylinder bore. Exhaust valve permits the burnt gases to escape from the cylinder bore at proper timing.

- Crank Pin: Hand over the power and motion to the crank shaft which come from piston through connecting rod.

- Governor: It controls the speed of engine at a different load by regulating fuel supply in diesel engine. In petrol engine, supplying the mixture of air-petrol and controlling the speed at various load condition.

- Cam Shaft: It takes driving force from crankshaft through gear train or chain and operates the inlet valve as well as exhaust valve with the help of cam followers, push rod and rocker arms.

- Fuel Pump: This device supply the petrol to the carburettor sucking from the fuel tank.

- Carburetor: It converts petrol in fine spray and mixes with air in proper ratio as per requirement of the engine.

- Fuel Injector: This device is used in diesel engine only and delivers fuel in fine spray under pressure.

- Spark Plug: These devices are used in petrol engine only and ignite the charge of fuel for combustion.

Working Principle of 4-Stroke Engine

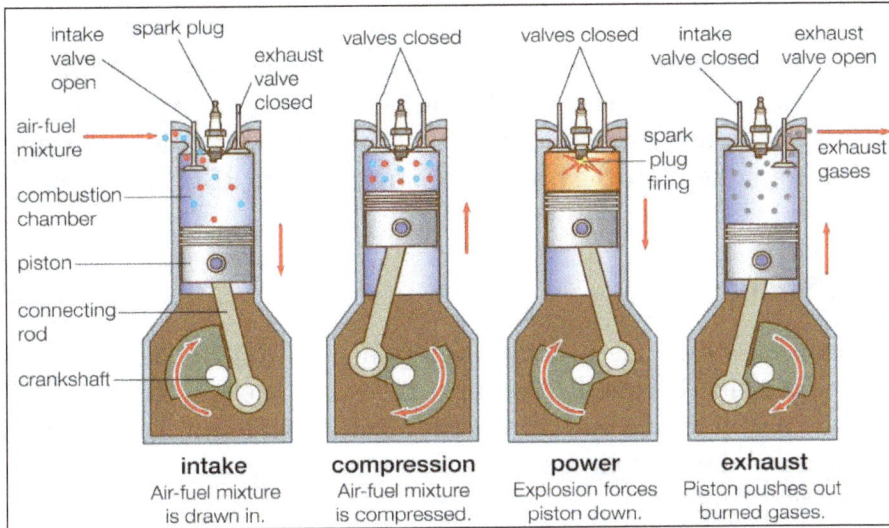

(a)Suction stroke (b) Compression stroke (c) Power stroke (d) Exhaust stroke.

It consists of the following four strokes:

- Suction stroke.

- Compression stroke.

- Power (or) Expansion stroke.

- Exhaust stroke.

1. Suction Stroke

At the beginning of the stroke the piston is at the top most position (TDC) and is ready to move downward. As the piston moves downwards vacuum will create inside the cylinder. Due to this vacuum air fuel mixture from the carburetor is sucked into the cylinder through inlet valves till the piston reaches bottom most position (BDC). During the suction stroke, exhaust valve remains in closed condition and inlet valve remains open. At the end of the suction stroke the inlet valve will be closed.

2. Compression Stroke

During the compression stroke both the inlet and exhaust values are in closed condition and the piston moves upward from BDC to compress the air fuel mixture. This process will continue till the piston reaches TDC as shown in figure (b). The compression ratio of engine varies from 6 to 8. The pressure at the end of compression is about 600 to 1200 Kg/m². The temperature at the end of the compression is 250 to 300°C. At the end of this stroke the mixture is ignited by spark plug. If leads to increase in pressure and temperature of the mixture instantaneously.

3. Power or Expansion Stroke

Both the pressure and temperature range of the ignited mixture are 1800 to 2000°C and 3000 to 4000 kN/m² respectively. During the expansion stroke both the values are remains closed. The rise in pressure of the mixture exerts an impulse on the piston and pushes it downward therefore the piston move from TDC to BDC. This stroke is known as power stroke. Refer figure (c).

4. Exhaust Stroke

During the exhaust stroke the piston moves from BDC to TDC the exhaust value is opened and inlet value is closed. The burnt gases are released through the exhaust value when the piston moves upward. As the piston reaches the TDC again the, inlet valves will open and the fresh air fuel mixture enters into the cylinder for the next cycle of operation. Refer figure (d).

It is obvious from the above operation only one power stroke is produced in each and every four stroke of the piston or two revolution of the crank-shaft. Hence it is termed as four stroke engine.

1.1.2 Valve and Port Timing Diagrams
1. Inlet Port

The Inlet Port opens at 35° to 50° prior to the TDC position which closes in equal amount after TDC position.

2. Exhaust Port

The exhaust port opens and closes at 35° − 70° before and after BDC Position respectively.

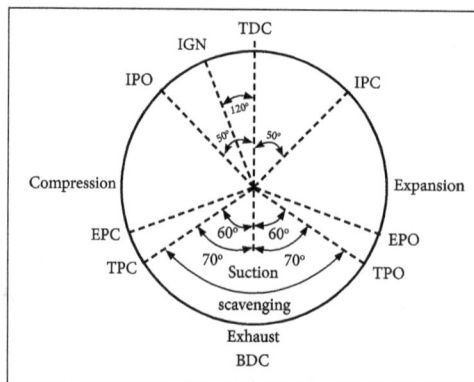

Port Timing Diagram.

- T.D.C - Top Dead Centre.
- I.P.O - Inlet Port Open.
- I.P.C - Inlet Port Close.
- T.P.O - Transfer Port Open.
- T.P.C - Transfer Part Close.
- E.P.O - Exhaust Port Open.
- E.P.C - Exhaust Port Close.
- B.D.C - Bottom Dead Centre.

3. Transfer Port

The transfer port opens at 35° to 60° in advance to the BDC position and closes at 35° to 60° after TDC position.

4. Ignition

Fuel injection valve opens at 10° to 15° before TDC position as the air requires some time to start ignition which closes at 15° to 20° after TDC position for better combustion. The scavenging period of petrol engines must not exceed above 70° whereas this period is large in case of diesel engines.

Advantages of Two-Stroke Engines Over Four Stroke Engines

- Construction is simple due to the absence of valves as the design of ports is much simpler and easy to manufacture.
- High mechanical efficiency due to the absence of cams, cam shaft and rockers etc., of the valve gear.

Disadvantages of Two-Stroke Engines Over Four Stroke Engines

- Thermal efficiency is low due to high compression ratio.
- There is a great wear and tear of moving parts.
- Two stroke engine produces great noise during exhaust stroke.

Theoretical Valve Timing Diagram

In theoretical valve timing diagram, inlet and exhaust valves open and close at both the dead centers. Similarly, all the processes are sharply completed at the TDC or BDC.

Figure shows theoretical valve timing diagram for four strokes S.I. Engines:

- IVO - Inlet Valve Open.

- IVC - Inlet Valve Close.

- IS - Ignition Starts.

- EVO - Exhaust Valve Open.

- EVC - Exhaust Valve Close.

- TDC - Top Dead Center.

- BDC - Bottom Dead Center.

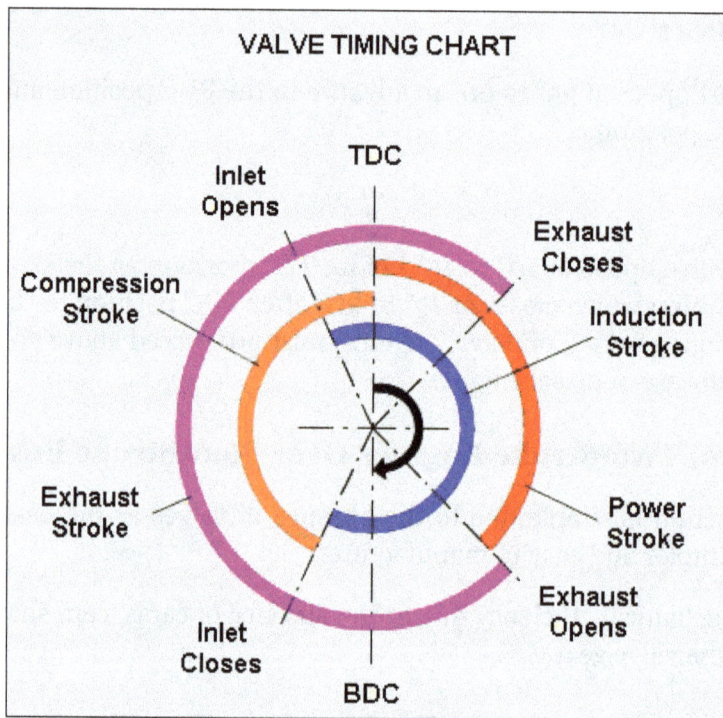

Valve Timing Diagram.

Actual Valve Timing Diagram

Figure shows actual valve timing diagram for four stroke S.I. engine. The inlet valve opens 10-30° before the TDC. The air-fuel mixture is sucked into the cylinder till the inlet Valve closes. The inlet valve closes 30-40° or even 60° after the BDC. The charge is compressed till the spark occurs. The spark is produced 20-40° before the TDC. It gives sufficient time for the fuel to burn. The pressure and temperature increase. The burnt gases are expanded till the exhaust valve opens.

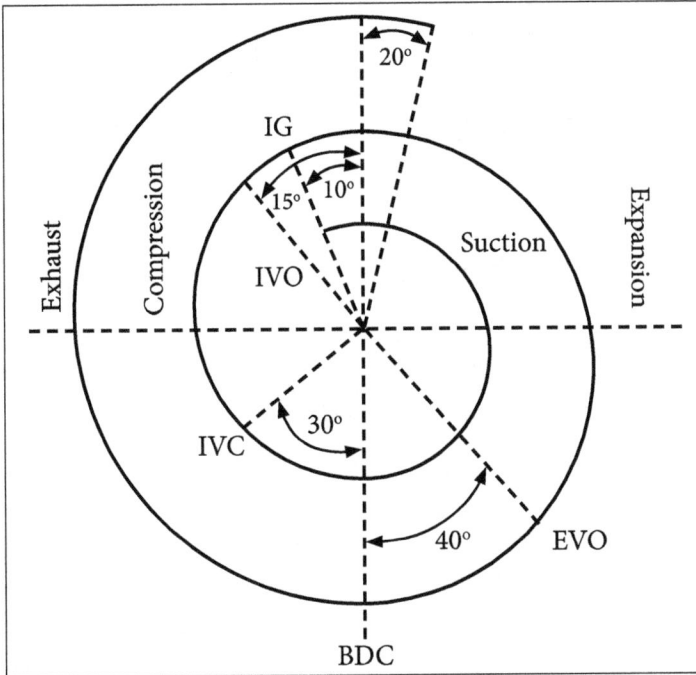

The exhaust valve opens 30-60° before the BDC. The exhaust gases are forced out from the cylinder till the exhaust valve closes. The exhaust valve closes 8-20° after the TDC. Before closing, again the inlet valve opens 10-30° before the TDC. The period between the IVO and EVC is known as valve overlap period. The angle between the inlet valve opening and exhaust valve closing is known as angle of valve overlap.

1.2 Air Standard Cycles

The air standard cycle is a cycle followed by a heat engine which uses air as the working medium. Since the air standard analysis is the simplest and most idealistic, such cycles are also called ideal cycles and the engine running on such cycles are called ideal engines. In order that the analysis is made as simple as possible, certain assumptions have to be made.

These assumptions result in an analysis that is far from correct for most actual combustion engine processes, but the analysis is of considerable value for indicating the upper limit of performance. The analysis is also a simple means for indicating the relative effects of principal variables of the cycle and the relative size of the apparatus.

The Carnot Cycle

This cycle is highest possible efficiency for any cycle. Figures show the P-V and T-s diagrams of the cycle.

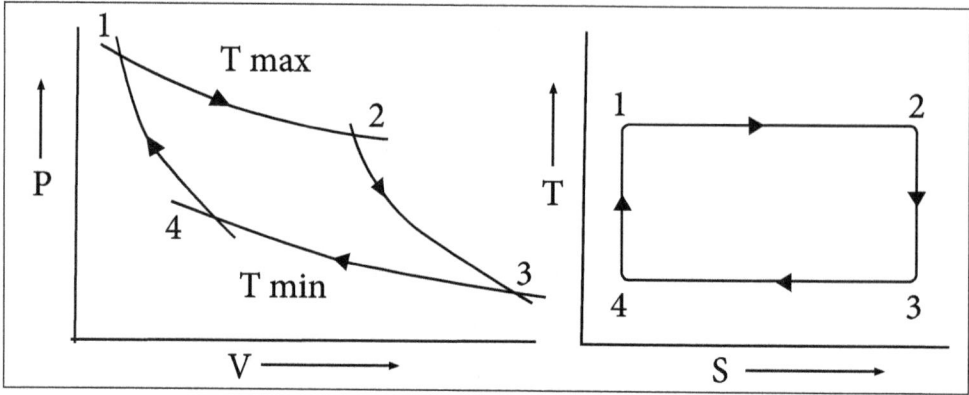

P-V Diagram of Carnot Cycle T-S Diagram of Carnot Cycle.

Assuming that the charge is introduced into the engine at point 1, it undergoes isentropic compression from 4 to 1. The temperature of the charge rises from T_{min} to T_{max}. At point 2, heat is added isothermally. This causes the air to expand, forcing the piston forward, hence doing work on the piston. At the point 3, the source of heat is removed at a constant temperature. At the point 4, a cold body is applied to the end of the cylinder and the piston reverses, therefore compressing the air isothermally; heat is rejected to the cold body. At the point 1, the cold body is removed and the charge is compressed isentropic ally till it reaches the temperature T_{max} once again. Hence, the heat addition and the rejection processes are isothermal while compression and expansion processes are isentropic.

From the theory of thermodynamics, per unit mass of charge,

Heat supplied from point 1 to 2 $= p_2 v_2 \ \text{In} \ \dfrac{V_2}{V_1}$

Heat rejected from point 3 to 4 $= p_3 v_3 \ \text{In} \ \dfrac{V_4}{V_3}$

Now,

$$p_2 \, V_2 = RT_{max}$$

And,

$$p_4 v_4 = RT_{min}$$

Since work done, per unit mass of charge, W = Heat supplied – heat rejected,

$$W = RT_{max} \ \text{In} \ \frac{V_3}{V_2} - RT_{min} \ \text{In} \ \frac{V_1}{V_4}$$

$$= R \ \text{In}(r)(T_{max} - T_{min})$$

We have assumed that the compression and expansion ratios are equal, that is $\dfrac{V_3}{V_2} = \dfrac{V_1}{V_4}$.

Heat supplied, $Q_s = RT_{max} \text{ In}(r)$

Hence, the thermal efficiency of the cycle is given by,

$$\eta_{th} = \frac{R \text{ In}(r)(T_{max} - T_{min})}{R \text{ In}(r) T_{max}}$$

$$= \frac{T_{max} - T_{min}}{T_{max}} \qquad ...(1)$$

From Equation 1, it is seen that the thermal efficiency of a Carnot cycle is only a function of maximum and minimum temperatures of the cycle. The efficiency will increase when the minimum temperature is as low as possible.

According to this Eq., the efficiency will be equal to 1 when the minimum temperature is zero, which happens to be the absolute zero temperature in the thermodynamic scale. This equation also indicates that for optimum efficiency, the cycle must operate between the limits of the highest and lowest possible temperatures. In other words, the engine should take in all the heat at as high a temperature as possible and should reject the heat at as low a temperature as possible.

To achieve the first condition, combustion should begin at the highest possible temperature, only then the irreversibility of the chemical reaction will be reduced. Moreover, the expansion should proceed to the lowest possible temperature in the cycle, in order to obtain the maximum amount of work. These conditions are the aims of all designers of modern heat engines. The conditions of heat rejection are governed, in practice, by the temperature of atmosphere.

It is not possible to construct an engine which will work on the Carnot cycle. Since, it would be necessary for the piston to move very slowly during the first part of the forward stroke so that it can follow an isothermal process.

During the remainder of the forward stroke, the piston would need to move very quickly as it has to follow an isentropic process. This variation in the speed of the piston could not be achieved in practice. Also, a very long piston stroke would produce only a small amount of work most of which will be absorbed by the friction of the moving parts of the engine.

1.2.1 Otto Cycle

The Otto cycle was first proposed by a Frenchman, Beau de Rochas in 1862, was first used on an engine built by a German, Nicholas A. Otto, in 1876. The cycle is also known as a constant volume or an explosion cycle. This is the equivalent air cycle for reciprocating piston engines using spark ignition. The below figure shows the P-V and T-S diagrams respectively.

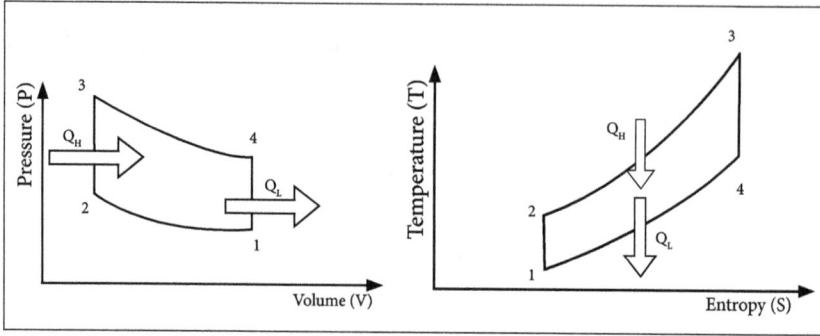

P-V Diagram of Otto Cycle and T-S Diagram of Otto Cycle.

At the start of the cycle, the cylinder has a mass M of air at the pressure and volume indicated at point (1). The piston is at its lowest position. It moves upward and the gas is being compressed isentropically to point (2). At this point, heat is added at constant volume which raises the pressure to point (3). The high pressure charge now expands isentropically, pushing the piston down on its expansion stroke to point (4) where the charge rejects heat at constant volume to the initial state, point 1.

The isothermal heat addition and rejection of the Carnot cycle is replaced by the constant volume processes which are, theoretically more possible, although in practice, even these processes are not possible.

The heat supplied, Q_s, per unit mass of charge, is given by,

$$C_v \left(T_3 - T_2 \right)$$

The heat rejected, Q_r per unit mass of charge is given by,

$$C_v \left(T_4 - T_1 \right)$$

And the thermal efficiency is given by,

$$\eta_{th} = 1 - \frac{\left(T_4 - T_1 \right)}{T_3 - T_2}$$

$$= 1 - \frac{T_1}{T_2} \left\{ \frac{\left(\dfrac{T_4}{T_1} - 1 \right)}{\left(\dfrac{T_3}{T_2} - 1 \right)} \right\} \qquad \ldots(1)$$

Now,

$$\frac{T_1}{T_2} = \left(\frac{V_2}{V_1} \right)^{\gamma-1} = \left(\frac{V_3}{V_4} \right)^{\gamma-1} = \frac{T_4}{T_3}$$

And Since,

$$\frac{T_1}{T_2} = \frac{T_4}{T_3}$$

We have,

$$\frac{T_4}{T_1} = \frac{T_3}{T_2}$$

Therefore, substituting in Eq. 1, assuming that r is the compression ratio V_1 / V_2, we get,

$$\eta_{th} = 1 - \frac{T_1}{T_2}$$

$$= 1 - \left(\frac{V_2}{V_1}\right)^{\gamma-1}$$

$$= 1 - \frac{1}{r^{\gamma-1}} \qquad \qquad ...(2)$$

In a true thermodynamic cycle, the term expansion ratio and compression ratio has the same meaning. But, in a real engine, these two ratios are not necessary be equal because of the valve timing and therefore the term expansion ratio is preferred. Equation 2 shows that the thermal efficiency of the theoretical Otto cycle increases with increase in compression ratio and specific heat ratio but is independent of the heat added (independent of load) and initial conditions of pressure, volume and temperature.

The below figure shown a plot of thermal efficiency versus compression ratio for Otto cycle. It is seen that the increase in efficiency is significant at the lower compression ratios.

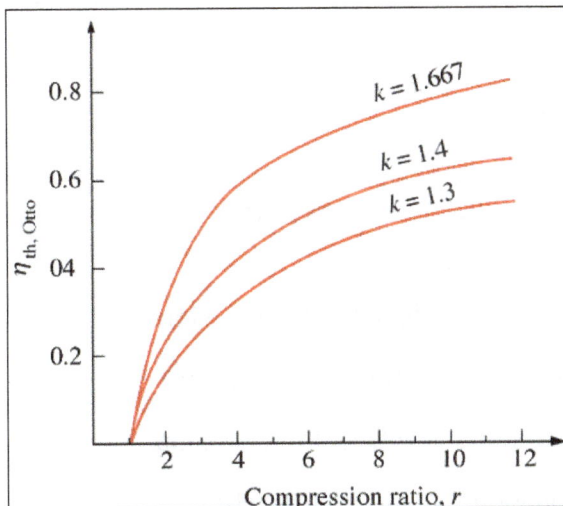

Variation of Efficiency with Compression Ratio.

1.2.2 Diesel Cycle

This cycle is a constant pressure cycle. This is believed to be the equivalent air cycle for reciprocating slow speed compression ignition engine. The P-V and T-s diagrams are shown in the below figure.

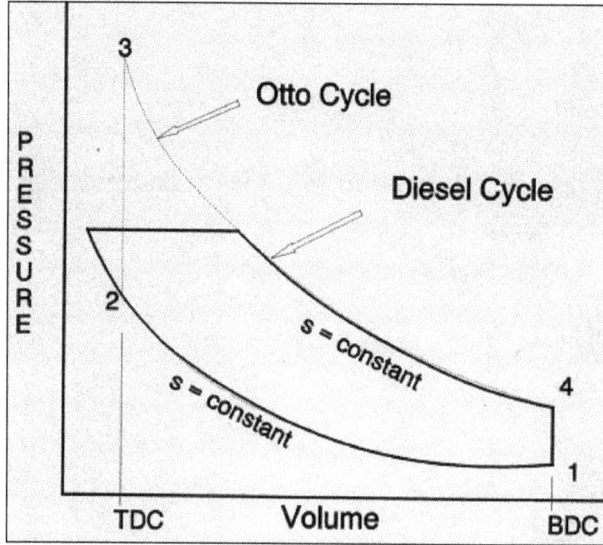

P-V Diagram of Diesel Cycle.

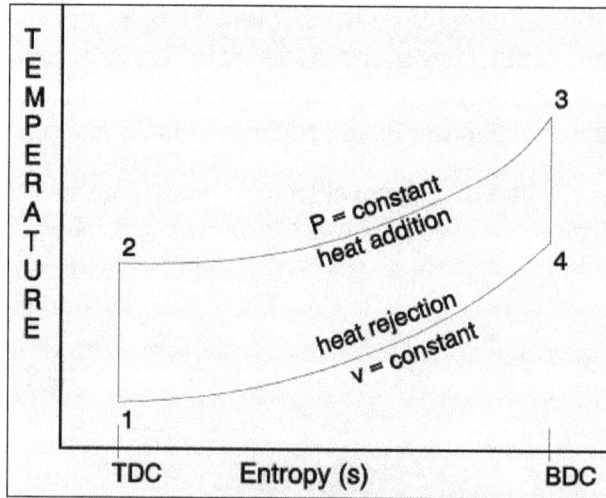

T-S Diagram of Diesel Cycle.

- Process 1-2: Reversible adiabatic Compression.

- Process 2-3: Constant pressure heat addition.

- Process 3-4: Reversible adiabatic Compression.

- Process 4-1: Constant volume heat rejection.

Consider 'm' kg of the working fluid. As the compression and expansion processes are reversible adiabatic processes, we can write it as follows,

Heat supplied $= m\, C_p\left(T_3 - T_2\right) = \left(h_3 - h_2\right)$

Heat rejected $= m C_v\left(T_4 - T_1\right) = \left(u_4 - u_1\right)$

Work done $= m\, C_p\left(T_3 - T_2\right) - m C_v\left(T_4 - T_1\right)$

Now, the thermal efficiency can be written as,

$$\eta_{th} = \frac{m C_p\left(T_3 - T_2\right) - m C_v\left(T_4 - T_1\right)}{m\, C_p\left(T_3 - T_2\right)}$$

$$= 1 - \frac{1}{\gamma}\left(\frac{T_4 - T_1}{T_3 - T_2}\right)$$

$$T_2 = T_1\, r^{\gamma-1}\ ;\ r = \frac{V_1}{V_2} = \frac{V_4}{V_2}$$

$$\frac{T_3}{T_2} = \frac{V_3}{V_2} = r_c = \text{cutoff ratio}$$

$$T_3 = r_c\, T_2 = r_c\, T_1\, r^{\gamma-1}$$

$$T_4 = T_3\left(\frac{V_3}{V_4}\right)^{\gamma-1} = T_3\left(\frac{V_4}{V_3}\right)^{\gamma-1}$$

$$= T_3\left(\frac{V_4}{V_2} \cdot \frac{V_2}{V_3}\right)^{1-\gamma} = T_3\left(\frac{r}{r_c}\right)^{1-\gamma}$$

$$= r_c\, T_1\, r^{\gamma-1}\left(\frac{r}{r_c}\right)^{1-\gamma}\ ;\ T_4 = r_c^{\gamma}\, T_1$$

$$\eta_{th} = 1 - \frac{1}{\gamma}\left\{\frac{r_c^{\gamma}\, T_1 - T_1}{r_c\, r^{\gamma-1}\, T_1 - r^{\gamma-1}\, T_1}\right\}$$

$$= 1 - r^{1-\gamma}\left\{\frac{r_c^{\gamma} - 1}{\gamma\left(r_c - 1\right)}\right\}$$

From the equation above, it is observed that, the thermal efficiency of the diesel engine may be increased by increasing the compression ratio, r, by decreasing the cut-off ratio, α_2 or by using gas with large value of γ. As the quantity $\left(r^{\gamma} - 1\right)/\gamma\left(r_p - 1\right)$ in above equation is always greater than unity, always the efficiency of a Diesel cycle is lower than that of an Otto cycle having the same compression ratio. However practical Diesel engines uses higher compression ratios compared to petrol engines.

1.3 Engine Systems: Fuel Injection and Carburetion

Battery Ignition System or Coil Ignition System

It is employed in petrol engines. The figure below shows the wiring diagram of a simple coil ignition system of a four cylinder engine. This system is used in automobiles.

Construction

It consists of a batter, ignition coil, condenser, contract breaker, distributor and spark plugs. Generally 6 or 12 volts battery is used. The ignition coil consists of two windings Primary and Secondary. The primary winding consists of thick wire with less root turns. The primary, winding is formed of 200-300 turns of thick wire of 20-gage to produce a resistance of about 1.5 ohms.

The secondary winding located inside the primary winding consists of about 21,000 tons of thin enameled wire of 38-40 gauges with sufficiently insulated to withstand high voltage. It is wound close to the core with one end connected to the secondary terminal and the other end rounded either to the metal case or the primary coil.

It prevents excess arcing and pitting of contact breaker points. The contact breaker is housed in the distributor itself, it makes and breaks the primary ignition circuit. The distributor distributes the high voltage to the respective. Spark plugs having regular intervals in the sequence of firing order of the engine.

(The sequence in which the firing or power occurs in a multi-cylinder engine is known as firing order. The firing order of a 4-cylinder in-line engine is 1-3-4-2 or 1-4-3-2. The firing order of a 6-cylinder in-line engine is 1-5-3-6-2-4).

The spark plug is fitted on the combustion chamber of the engine. It produces spark to ignite the fuel-air mixture. The motor of the distributor and contact breaker cam are driven by the engine. There are two circuits in this system. One is the primary circuit. It consists of battery, primary coil of the ignition coil. Condenser and contact breaker. The other circuit is the secondary circuit. It consists of secondary coil, distributor and spark plug.

Working

The ignition switch is switch on and the engine is cranked. The cranking of the engine opens and closes the contact breaker points through a cam.

When Contact Breaker Points are Closed

- The current flows from the battery to the contact breaker points through the switch and primary winding and then returns to battery through the earth.

- This current build up a magnetic field in the primary winding of the ignition coil.

- When the primary current is at the higher peaking contact breaker points will be opened by the cam.

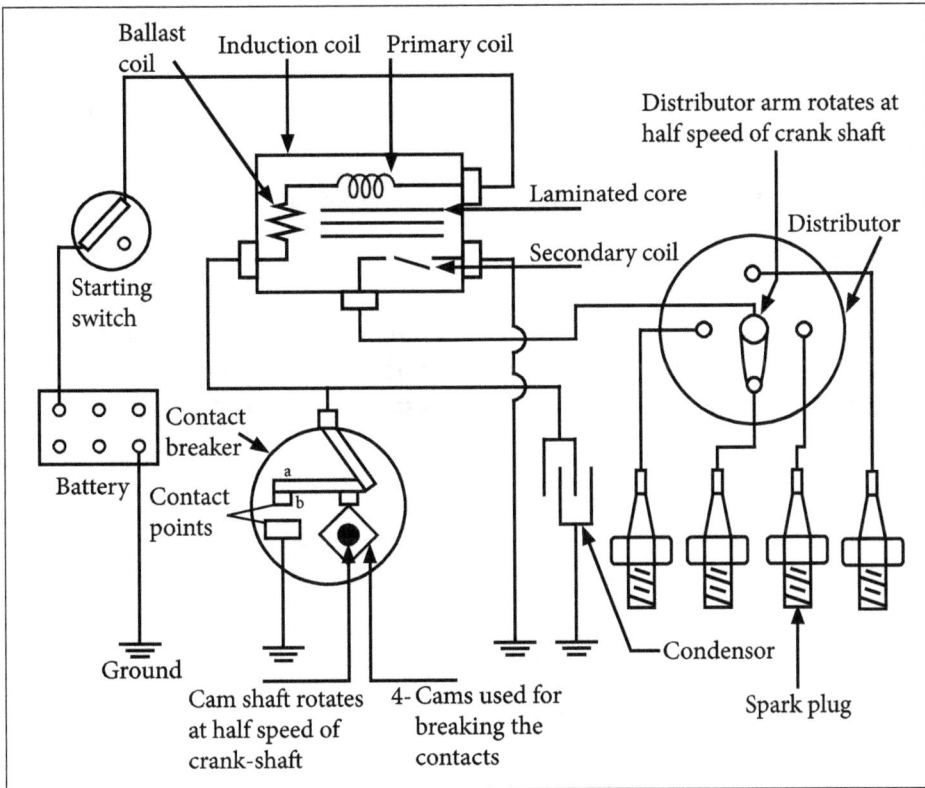

When the Contact Breaker Points are Opened

- The magnetic field set up in the primary winding is suddenly collapsed.

- A high voltage (15000 volts) is generated in the secondary winding of ignition coil.

- This high voltage is directed to the rotor of the distributor.

- The rotor directs this high voltage to the individual spark plugs in the sequence of the firing or devotes the engine.

- This high voltage tries to cross the spark plug gap (0.45 to 0.6 mm) and spark is produced. The spark ignites the fuel-air mixture.

Advantages

- It provides better sparks at low speeds of the engine during. Starting and idling due to availability of maximum current-thought the engine speed range.

- The initial cost is low as compares with magneto ignition system.

- The maintenance cost is negligible except battery.

- Spark efficiency remains unaffected by various positions of the timing control mechanism.

Disadvantages

- Frequency battery down occurs when the engine is not in use continuously. This causes starting trouble.

- The weight is greater that magneto ignition system.

- Wiring mechanism is more complicated.

In this system, the battery is replaced with a magneto. Figure, shows the wiring diagram of a magneto ignition system. It consists of a switch, magneto, contact breaker, condenser, distributor and spark plugs. This system is used in two wheeler like motor cycles, scooters etc.

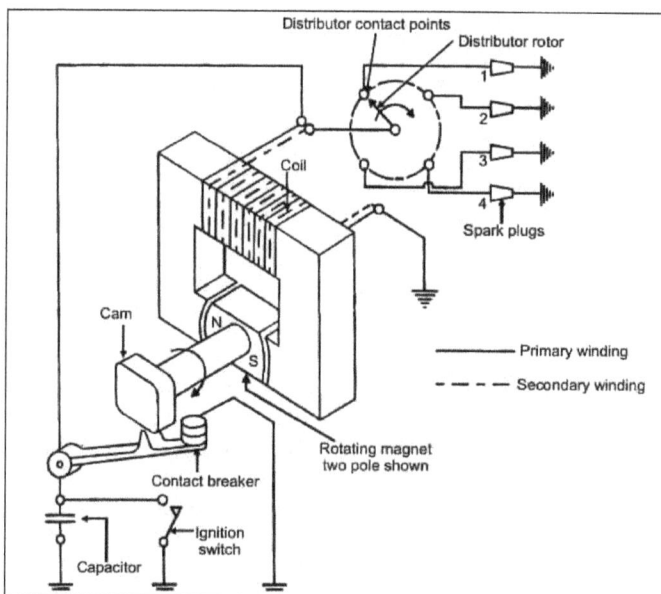

Construction

The magneto ignition system consists of a rotating magnet assembly driven by a engine and a fixed armature. The armature consists of primary and secondary windings. The primary circuit consists of a primary winding condenser and contact breaker. The secondary circuit consists of a secondary windings, distributor and spark plugs.

When the Contact Breaker Points are Closed

- The current flows in the primary closed.

- This produces a magnetic field int he primary winding.

- When the primary current is at the highest peak, the contact breaker points will be opened by the cam.

When the Contact Breaker Points are Opened

- There is a break in the primary circuit.

- The magnetic field in the primary winding is suddenly collapsed.

- A high voltage (15000 volts) is generated in the secondary winding.

- This high voltage is distributed to the respective spark plugs through the rotor of the distributor.

- The high voltage tries to cross the spark plug gap and a spark is produced in the gap. This speak ignites the fuel-air mixture in the engine cylinder.

Advantages

- It has no maintenance problem like coil ignition (i.e., for battery). So, it is more reliable.

- When the speed increases, it provides better intensity of spark and thus provides better combustion as compared to battery coil ignition system.

- Less space is required as compared to battery ignition system.

- It is very light in weight and compact in size.

Disadvantages

- Initial cost is very high as compared to coil ignition system.

- Minimum 75rpm is necessary to start the engine.

- For higher power engines, some other devices are necessary to start ignition.

Carburetor

The main components of simple carburetor are: float chamber, float, nozzle, venture, inlet valve and metering jet. In the float chamber, a constant level of petrol is maintained by the float and a needle valve. The float chamber is ventilated to atmosphere. This is used to maintain atmospheric pressure inside the chamber.

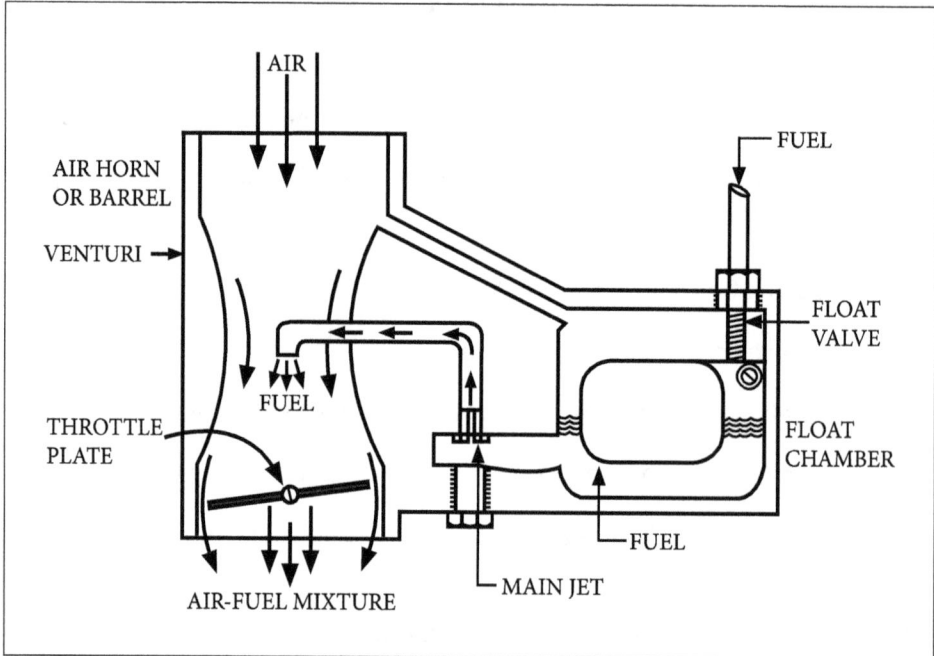

Carburetor

The float which is normally a metallic hollow cylinder rise and close the inlet valve as the fuel level in the float chamber increases to certain level.

The mixing chamber contains venture, nozzle and throttle valve, the venturi tube is fitted with the inlet manifold. This tube is narrow opening called venturi. A nozzle provided jet below the center of this venturi. The nozzle keeps the same level of petrol as the level in the float chamber. The mixing chamber has two butterfly valves. One is to allow air in to the mixing chamber known as choke valve. The other is to allow air fuel mixture to the engine known as throttle valve.

Working

During the suction stroke, vacuum is created inside the cylinder. This causes pressure difference between the cylinder and outside the carburetor. Due to this, the atmospheric air enters in to the carburetor. The air flows through the venturi. This produces the partial vacuum at the tip of the nozzle. Because of this vacuum, the fuel comes out from the nozzle in the form of fine spray. These fine fuel partial mix with incoming air to form air-fuel mixture. Thus, it gives a homogeneous mixture of air fuel to the engine.

1.3.1 Ignition, Cooling and Lubrication

Comparison Between Battery Ignition and Magneto Ignition System

S. No	Battery Ignition	Magneto ignition
1	Battery is a basic requirement.	No battery needed.
2	A good spark is available at low speed also.	During starting the quality of spark is poor due to slow speed.
3	Occupies more space.	Very much compact.
4	Mostly employed in car and bus for which it is required to crank the engine.	Used on motorcycles, scooters, etc.
5	Battery maintenance is required.	No battery maintenance problems.
6	Battery supplies current in primary circuit.	Magneto produces the required current for primary circuit.
7	Recharging is a must in case battery gets discharged.	No such arrangement required.

Lubrication

Lubrication plays a main role in life expectancy of an engine. Without oil, an engine would succumb to overheating and seize very quickly. Lubricants help in mitigating this problem, and if properly monitored and maintained, can extend the life of your motor.

The main functions of the lubrication system is to:

- Reduce friction and wear between the parts having relative motion.

- Cool the surface by carrying away heat generated due to friction.

- Seal a space adjoining the surface. Such as piston rings and cylinder lines.

- Clean the surface by carrying away the carbon and metal particles caused by wear.

- Absorb shock between bearings and other parts and consequently reduce noise.

All the parts of the engine are efficiency lubricated. The minute gap between the sliding surfaces can be lubricated since the oil is supplied under pressure.

Example: Crank shaft main bearings, can shaft main bearing, big end bearings of the connecting rod etc.

Various lubrication systems used for I.C. engines maybe classified as:

- Wet sump lubrication system.

- Dry sump lubrication system.

- Mist lubrication system.

1. Wet Sump Lubrication System

These systems employ a large capacity oil sump at the base of crank chamber, from which the oil is drawn by a low pressure oil pump and delivered to various parts. Oil there gradually returns back to the sump after serving the purpose.

Splash System

This system is used on some small four-stroke stationary engines. In this case the caps on the big ends bearings of connecting rods are provided with scoops which, when the connecting rod is in the lowest position, just dip into oil troughs and thus direct the oil through holes in the caps to the big end bearings. Due to splash of oil it reaches the lower portion of the cylinder walls, crankshaft and other parts requiring lubrication. Surplus oil eventually flows back to the oil sump. Oil level in the troughs is maintained by means of an oil pump which takes oil from sump, through a filter. Splash system is suitable for low and medium speed engines having moderate bearing load pressures. For high performance engines, which normally operate at high bearing pressures and rubbing speeds this system does not serve the purpose.

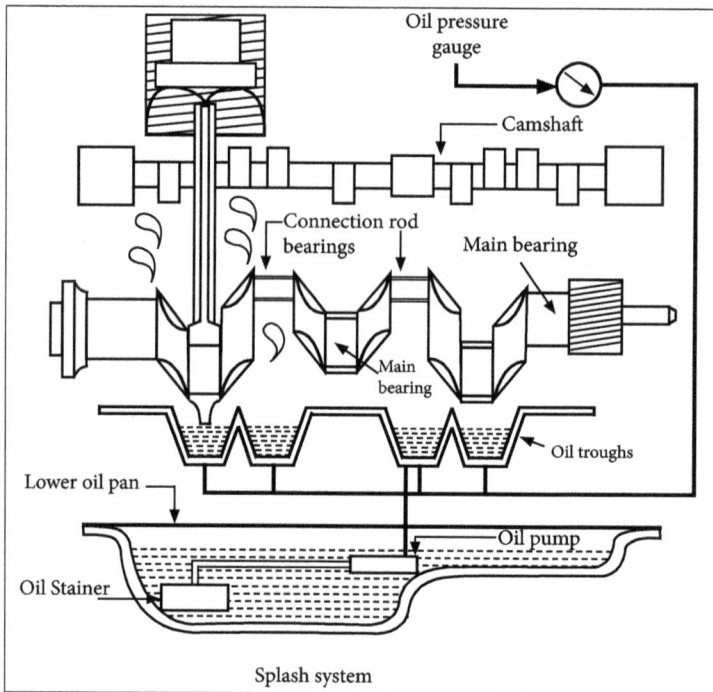

Splash system

Cooling Systems

When the air-fuel mixture is ignified and combustion takes place at about 2500°C for producing power inside an engine the temperature of the cylinder, cylinder head, piston and valve, continuous to rise when the engine runs. If these parts are not cooled by some means then they likely to get damaged and even melted. The piston may cease inside the cylinder. To prevent thus, the temperature of the parts around the combustion, chamber is maintained as 200°C to 250°C. Too much of cooling will lower the thermal efficiency of the engine. Hence the purpose of cooling is to keep the engine at its most efficient operating temperature at all engine speeds and all driving conditions.

Water Cooling System

In water-cooling, water is used for cooling the engine by circulating it through water jackets around each combustion chamber cylinder, cylinder head, valve and valve sheet. By absorbing heat, water will become hot. When it is again passed through radiator, it will be cooled by air blast due to forward motion of the vehicle as well as of this engine to absorb heat.

There are two systems of water-cooling system. They are:

- Thermo syphon system.

- Pump circulation system.

1. Thermo Syphon System

The principle of hot water going up and cold Water coming down due to difference in density is used here. There is no pump to circulate water. The light hot water from the engine goes to the top of the radiator by itself and gets cooled by the surrounding air and hence goes down to bottom of radiator and again goes to engine cylinder as shown in figure.

It is simple, cheap but cooling is slow. Water should be maintained to correct level at all time.

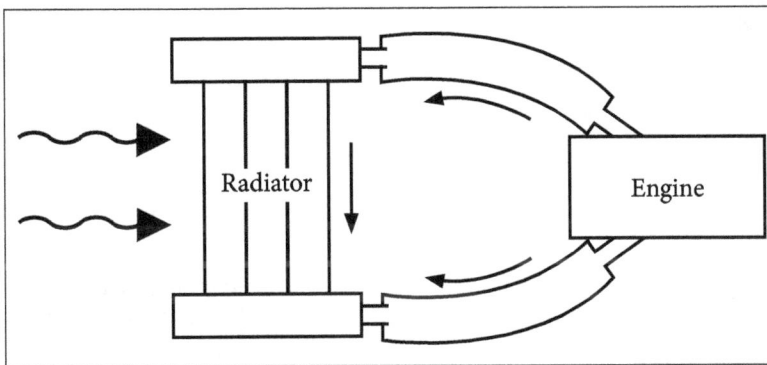

Thermo syphon system of cooling.

2. Pump Circulation System

To make the thermo syphon system more effective and improve water circulation, a water pump is introduced as shown in figure which is driven by a V-belt from a pulley on the engine crank shaft. This is called pump circulation system.

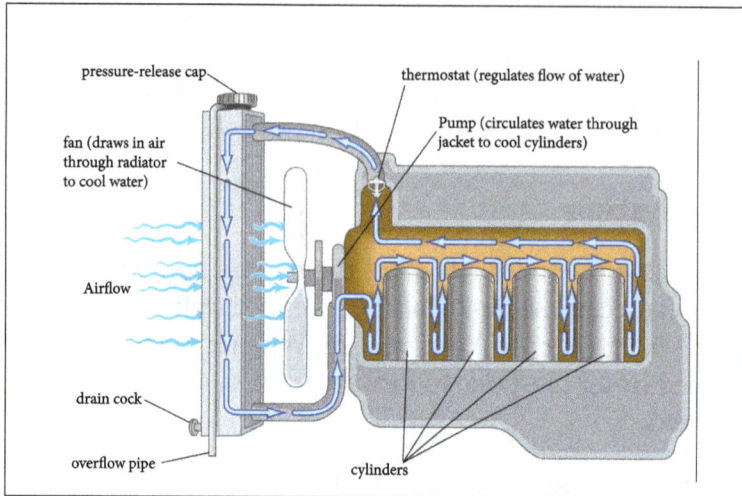

Water-cooling system for 4-cylinder engine.

The water-cooling arrangement for a 4 cylinder engine is shown in figure. When the hot water in engine passes through radiator tubes from upper tank to lower tank, it is exposed to large amount of airflow and gets sufficiently cooled. Then it is pumped to cylinder jackets by the water pump. The automatic thermostatic valve is used to regulate the circulation of water so that very cold water will become hot in short time to improve efficiency of the engine.

Performance Calculation

The power developed by an engine at the output shaft is called the brake power,

$$B_p = \frac{2\pi NT}{60} \text{ in W}$$

N = Speed in rpm

T = torque in N-m

Problems

1. During the trial (60 minutes) on a single cylinder oil engine having cylinder diameter 300 mm, stroke 450 mm and the working on four stroke cycle, the following observations were made:

- Total fuel used: 9.6 liters.

- C.V. of fuel: 45000 kJ/kg.

- Total No. of the Revolutions: 12624.

- Gross IMEP: 7.24 bar.

- Pumping TMEP: 0.34 bar.

- Net load on the brake: 3150N.

- Diameter of brake wheel drum: 1.78m.

- Diameter of the rope: 40 mm.

- Cooling water circulated: 545 liters.

- Rise in temperature of the Cooling water: 25°C.

- Specific gravity of oil: 0.8.

Let us determine the indicated power, brake power and mechanical efficiency.

Solution:

Given data:

$k = 1$; $D = 1.78$ m; $W = 3150$ N; $Dc = 0.3$m; $L = 0.45$m.

Mean effective pressure,

$$Pm = Pmg - Pmp$$

$$= 7.24 - 0.34 = 6.9 \text{ bar}$$

Pm   6.9 bar

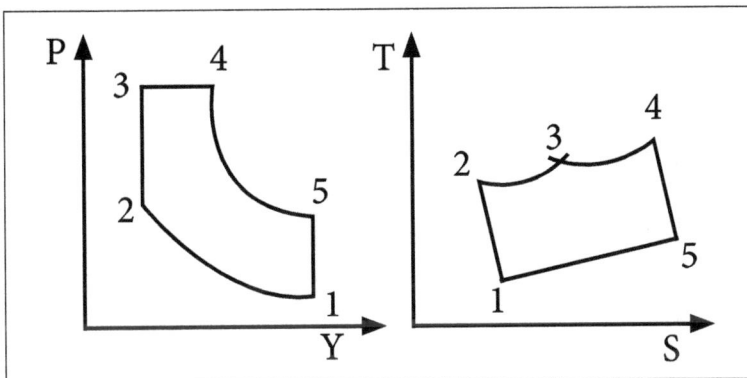

From process 1-2,

$$\frac{T_1}{T_2} = \frac{1}{(r)^{\gamma-1}} = \frac{1}{(12)^{1.4-1.0}} = \frac{1}{(12)^{0.4}}$$

$$\Rightarrow \frac{T_2}{T_1} = (12)^{0.4}$$

$$T_2 = (12)^{0.4} \times 300 = 810.576 \, k$$

$$T_2 = 810.576 \, k$$

$$\frac{P_2}{P_1} = (r)^\gamma = (12)^{1.4}$$

$$P_2 = (12)^{1.4} \times 1 \times 10^5 = 32.42 \times 10^5 \, N/m^2$$

$$P_2 = 32.42 \times 10^5 \, N/m^2$$

Pressure ratio,

$$r_P = \frac{P_3}{P_2} = \frac{70 \times 10^5}{32.42 \times 10^5} = 2.159$$

$$r_P = 2.159$$

Efficiency of cycle,

$$\eta_{dual} = 1 - \frac{t}{12^{0.4} \left[\dfrac{2.159 \times 1.33^{1.4} - 1}{(2.159 - 1) + (1.4 \times 2.139 \times 0.33)} \right]}$$

$$\eta_{dual} = 61.917 \%$$

Stroke volume,

$$\eta_{dual} = 61.917 \%$$

$$V_s = \frac{\pi}{4} \times d^2 \times L$$

$$= \frac{\pi}{4} \times 0.25^2 \times 0.3$$

$$V_s = 0.0147 \, m^3$$

Wkt,

$$\gamma = \frac{V_S + V_C}{V_C}$$

$$V_C = \frac{V_S}{\gamma - 1} = \frac{0.0147}{12 - 1}$$

2-3 is constant volume process,

$$V_c = V_2 = 1.338 \times 10^{-3} \, m^3$$

$$\Rightarrow V_2 = V_3 = 1.338 \times 10^{-3} \, m^3$$

$$\frac{P_2 V_2}{T_2} = \frac{P_3 V_3}{T_3}$$

$$T_3 = \frac{P_3}{P_2} \times T_2 = r_p \times T_2$$

$$T_3 = 1750.03\,k$$

We know that,

$$\rho = \frac{V_4}{V_3} \Rightarrow V_4 = e \times V_3 = 1.33 \times 1.328 \times 10^{-3}$$

$$V_4 = 1.779 \times 10^{-3} \ m^3$$

From process $3-4$; $(P = C)$

$$\frac{P/3 \ V_3}{T_3} = \frac{P/4 V_4}{T_4}$$

$$T_4 = \frac{V_4}{V_3} \times T_3 = \rho \times T_3 = 1.33 \times 1750.03$$

$$T_4 = 2327.54 \ K$$

$$\text{Speed } N = \frac{12624}{60} = 210.4 \ rpm$$

For four stroke engine $n = \dfrac{N}{2}$.

Indicated power,

$$I.P = \frac{P_m \ LA \times n \times k}{60}$$

$$= \frac{6.9 \times 10^5 \times 0.45 \times \dfrac{\pi}{4} \times 0.3^2 \times \dfrac{210.4}{2} \times 1}{60}$$

$$I.P = 38.428 \ kw$$

Brake power,

$$B.P = \frac{(W - S)\pi \ D_6 \ N}{60 \times 1000}$$

$$= \frac{3150 \times \pi \times 1.78 \times 210.4}{60 \times 1000}$$

$$B.P = 61.77 \ kW$$

2. In a constant speed compression ignition engine operating on four-stroke cycle and fitted with band brake, the following observations were taken:

- Brake wheel diameter 60 cm.
- Band thickness5 mm.
- Speed450 rpm.
- Load on band210 N.
- Spring balance reading30 N.
- Area of indicator diagram4.15 cm³.
- Length of indicator diagram6.25 cm³.
- Spring NO. 11, i.e11 bar/cm.
- Bore10 cm.
- Stroke15 cm.
- Specific fuel compression0.3 kg/kW-hr.
- Heating value of fuel41800 kJ/kg.

Let us determine the brake power, indicated power and mechanical efficiency.

Solution:

Given data:

$$D = 60 \text{ cm} = 0.6 \text{ m}$$

$$N = 450 \text{ rpm} = 7.5 \text{ rps}$$

$$W_1 = 210 \text{ N} = 0.21 \text{ kN}$$

$$W_2 = 30 \text{ N}$$

$$A = 41.5 \text{ cm}^2$$

$$L = 6.25 \text{ cm}$$

$$S = 11 \text{ bar/cm}$$

$$d = 10 \text{ cm} = 0.1 \text{ m}$$

$$I = 15 \text{ cm} = 0.15 \text{ m}$$

$m_f = 0.38$ kg/kw hr

$C_v = 41800$ kJ/kg

To find:

1. BP

2. IP

3. η_{mech}

Brake Power, BP = 2 NWR)

Net load on brake drum, $W = W_1 - W_2$
$$=210 - 30$$
$$= 180 \text{ N}$$
$$= 0.18 \text{ kN}$$

Effective brake radius,

$$R = \frac{D+d}{2} = \frac{0.6 + \left(5 \times 10^{-3}\right)}{2}$$

R = 0.3025 m

Break Power, BP = 2 N WR
$$= 2 \times 7.5 \times 0.18 \times 0.3025$$

BP = 2.56589 kW

Indicated Power, IP = P_m L A N

Indicated mean effective pressure Pm,

$$P_m = \frac{AS}{L}$$

$P_m = 730.4$ kN/m²

IP = P_m L A N

$$a = \frac{\pi}{4} d^2 = \frac{\pi}{4} \times 0.1^2$$

$a = 7.853 \times 10^{-3}$ m²

$$IP = P_m \, L \, A \, N$$

$$= 730.4 \times 0.15 \times 7.853 \times 103 \times 4 \times 1$$

$$I\,p = 13.767 \text{ kW}$$

Mechanical Efficiency,

$$\eta_{mech} = \frac{BP}{IP} = \frac{2.56589}{13.767} = 0.18638$$

1.3.2 Engine Performance Evaluation

The important performance parameters of I.C. engines are as follows:

- Friction Power.
- Indicated Power.
- Brake Power.
- Specific Fuel Consumption.
- Air Fuel ratio.
- Thermal Efficiency.
- Mechanical Efficiency.
- Volumetric Efficiency.
- Exhaust gas emissions.
- Noise.

The performance of an engine is usually studied by heat balance sheet. The main components of the heat balance are:

- Heat rejected to the cooling medium.
- Heat equivalent to the effective (brake) work of the engine.
- Unaccounted losses: The unaccounted losses include the radiation losses from the various parts of the engine and heat lost due to incomplete combustion.
- Heat carried away from the engine with the exhaust gases.

The friction loss is not shown as a separate item to the heat balance-sheet as the friction loss ultimately reappears as heat in cooling water, exhaust and radiation.

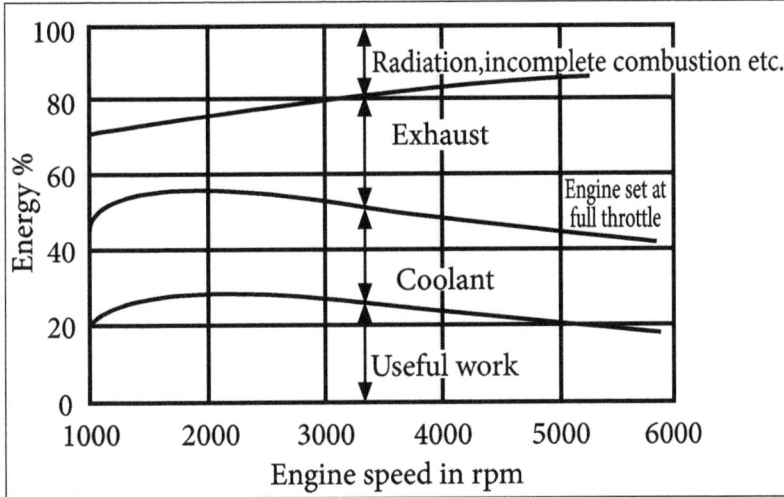

Heat Balance vs. Speed for a Petrol Engine at Full Throttle.

The performance of SI engine at constant speed and variable loads is different from the performance at full throttle and variable speed. Figure shows the heat balance of SI engine at constant speed and variable load. The load is varied by altering the throttle and the speed is kept constant by resetting the dynamometer.

Closing the throttle reduces the pressure inside the cylinders but the temperature is affected very little because the air/fuel ratio is substantially constant, and the gas temperatures throughout the cycle are high. This results in high loss to coolant at low engine load. This is reason of poor part load thermal efficiency of the SI engine compared with the CI engine:

- The loss to coolant is about 60 percent at low loads and 30 percent at full load.

- At low loads the efficiency is about 10 percent rising to about 25 percent at full load.

- Percentage loss to radiation increases from about 7% at loads or 20% at full load.

- The exhaust temperature rises very slowly with load and as mass flow rate of exhaust gas is reduced because the mass flow rate of fuel into the engine is reduced, the percentage loss to exhaust remains nearly constant.

The Performance of a CI Engine

The performance of a CI engine at constant speed variable load is shown in Figure:

- As the efficiency of the CI engine is more than the SI engine the total losses are less. The coolant loss is more at low loads and radiation, etc. losses are more at high loads.

- The b mep, bp and torque directly increase with load, as shown in Figure. Unlike the SI engine bhp and b mep are continuously raising curves and are limited only by the load.

The lowest brake specific fuel consumption and hence the maximum efficiency occurs at about 80 percent of the full load. The figure shows the performance curves of variable speed GM 7850 cc. four cycle V-6 Toro-flow diesel engine.

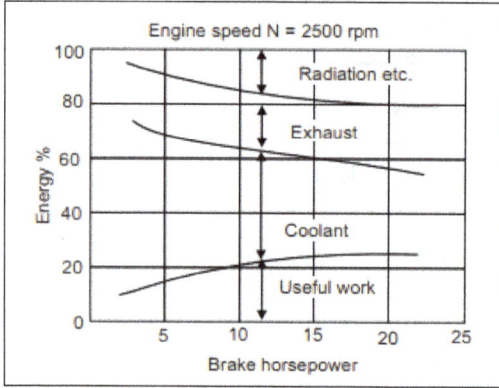

Heat Balance Vs. Load for a Petrol Engine   Heat Balance Vs. Load for a CI Engine.

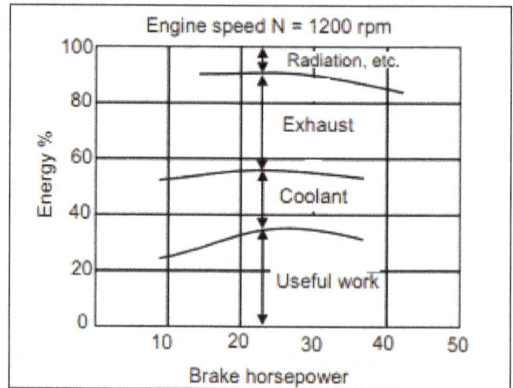

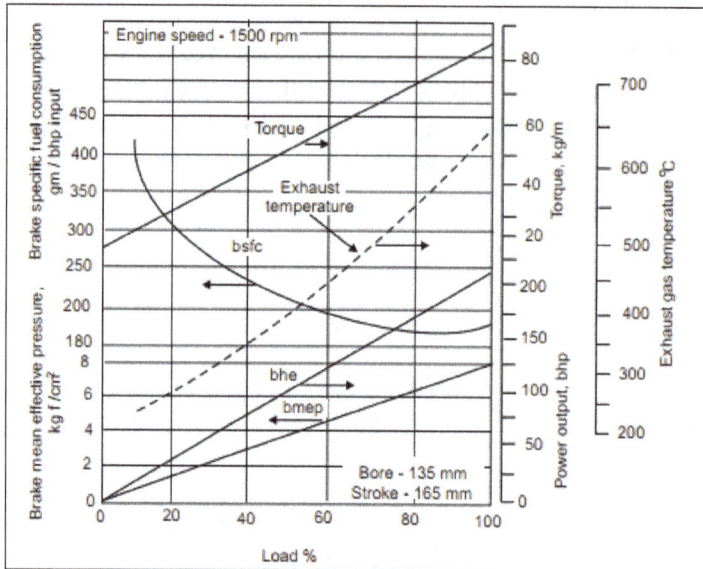

Performance Curves of a Six Cylinder Four-stroke Cycle
Automotive Type CI Engine at Constant Speed.

Problem

1. A gasoline engine works on the Otto cycle. It consumes 8 liter of gasoline per hour and develops the power at the rate of 25 kW. The specific gravity of gasoline is 0.8 and its calorific value is 44000 kJ/kg. Let us now determine the indicated thermal efficiency of the engine.

Solution:

Heat liberated at the input,

$$= m\,C_v$$

$$= 8 \times \frac{0.8}{60 \times 60}$$

$$= \frac{6.4}{3600}$$

Power in the input $= \frac{6.4}{3600} \times 44000$ kW

$$\eta_{ith} = \frac{\text{Output power}}{\text{Input power}}$$

$$= \frac{25}{\dfrac{6.4 \times 44000}{3600}}$$

$$= \frac{25 \times 3600}{44000} = 0.3196$$

or,

$$= 31.96\%$$

2. A single cylinder engine that is operating at 2000 rpm develops a torque of 8 N-m. The indicated power of engine is 2.0 kW. Find the loss due to friction as the percentage of brake power.

Solution:

$$\text{Brake power} = \frac{2\pi NT}{60000} = \frac{2 \times \pi \times 2000 \times 8}{60000}$$
$$= 1.6746 \text{ kW}$$

Friction power $= 2.0 - 1.6746$
$$= 0.3253$$

$$\% \text{ loss} = \frac{0.3253}{2} \times 100$$

$$\% \text{loss} = 16.2667\ \%$$

3. A six-cylinder, gasoline engine operates on the four-stroke cycle. The bore of each cylinder is 80 mm and the stroke is 100 mm. The clearance volume per cylinder is 70 cc. At the speed of 4100 rpm, the fuel consumption is 5.5 gm/sec. [or 19.8 kg/hr.) and the torque developed is 160 Nm.

Let us calculate:

- The brake mean effective pressure.
- Brake power.
- Brake thermal efficiency if the calorific value of the fuel is 44000 kJ/kg.
- The relative efficiency on a brake power basis assuming the engine works on the constant volume cycle r = 1.4 for air.

Solution:

$$bp = \frac{2\pi \; NT}{60000} = \frac{2 \times \pi \times 4100 \times 160}{60000} = 68.66$$

$$P_{bm} = \frac{bp \times 6000}{LA \, n \, K}$$

$$= \frac{68.66 \times 60000}{0.1 \times \frac{\pi}{4} \times (0.08)^2 \times \frac{4100}{2} \times 6}$$

$$= 6.66 \times 10^5 \; P_a$$

$$P_{bm} = 6.66 \text{ bar}$$

$$\eta_{bth} = \frac{bp}{m_f \times C_v} = \frac{68.66 \times 3600}{19.8 \times 43000} \times 100 = 29.03 \text{ \%}$$

Compression ratio, $r = \dfrac{V_s + V_d}{V_d}$

$$V_s = \frac{\pi}{4} D^2 \, L = \frac{\pi}{4} \times 8^2 \times 10 = 502.65 \, cc$$

$$r = \frac{502.65 + 70}{70}$$

$$r = 8.18$$

Air standard efficiency, $\eta_{otto} = 1 - \dfrac{1}{(8.18)^{0.4}} = 1 - \dfrac{1}{2.3179} = 0.56858$

Relative efficiency, $\eta_{rel} = \dfrac{0.2903}{0.568} \times 100 = 51.109\ \%$

$$\eta_{bth} = \dfrac{bp}{m_f \times C_v}$$

$$= \dfrac{119.82 \times 60}{\dfrac{4.4}{10} \times 44000}$$

$$\eta_{bth} = 37.134\ \%$$

Volume flow rate of air at intake condition,

$$a = \dfrac{6 \times 287 \times 300}{1 \times 10^5} = 5.17\,m^3/min$$

Swept volume per minute,

$$V_s = \dfrac{\pi}{4} D^2 L n K$$

$$= \dfrac{\pi}{4} \times (0.1)^2 \times 0.9 \times \dfrac{4500}{2} \times 9$$

$$= 127.17\,m^3/min$$

Volumetric efficiency, $\eta_v = \dfrac{5.17}{127.17} \times 100$

$$\eta_v = 4.654\ \%$$

Air – fuel ratio, $\dfrac{A}{F} = \dfrac{6.0}{0.44} = 13.64$

4. A six-cylinder, four-stroke engine gasoline engine that has a bore of 90 mm, stroke of 100 mm and has a compression ratio 8. The relative efficiency is 60%. When the indicated specific fuel consumption is 3009 g/kWh.

Estimate the:

- The calorific value of fuel.

- Corresponding fuel consumption given that i mep is 8.5 bar and speed is 2500 rpm.

Solution:

Air standard efficiency $=1-\dfrac{1}{r^{r-1}}=1-\dfrac{1}{8^{0.4}}=0.5647$

Relative efficiency $=\dfrac{\text{Thermal efficiency}}{\text{Air}-\text{standard efficiency}}$

Indicated thermal efficiency $=0.6\times0.5647=0.3388$

$$\eta_{ith} =\dfrac{1}{i_{sfc} \times C_v}$$

$$=\dfrac{1}{\eta_{ith} \times i_{sfc}} =\dfrac{3600}{0.3 \times 0.3388}$$

$$C_v =35417.035 \text{ kJ / kg}$$

$$IP =\dfrac{P_{im} L A N}{60000}$$

$$=\dfrac{8.5 \times10^5 \times0.1\times \dfrac{\pi}{4}\times0.09^2 \times\dfrac{2500}{2}\times6}{60000} =67.6\,\text{kW}$$

Fuel consumption $= i_{sfc} \times IP = 0.3 \times 67.6$

$$IP =20.28 \text{ kg/h}$$

5. Find air-fuel ratio of a 4-stroke, 1 cylinder, air cooled engine with fuel consumption time for 10 cc as 20.0 sec. The air consumption time for 0.1 m³ as 16.3 sec. The load is 16 kg at speed of 3000 rpm. Also find brake specific fuel consumption in g/kWh and the thermal brake efficiency. Consider the density of air as 1.175 kg/m³ and specific gravity of the fuel to be 0.7. The lower heating value of fuel is 44 MJ/kg and the dynamometer constant is 5000.

Solution:

Air consumption $=\dfrac{0.1}{16.3} \times1.175 =7.21 \times10^{-3} \text{ kg / s}$

Fuel consumption $=\dfrac{10}{20}\times0.7 \times\dfrac{1}{1000}=0.35 \times10^{-3} \text{ kg / s}$

Air-fuel ratio $=\dfrac{7.21 \times10^{-3}}{0.35 \times10^{-3}}=20.6$

Power output (P) $= \dfrac{\text{WN}}{\text{Dynamometer constant}}$

$$= \dfrac{16 \times 3000}{5000} = 9.6\,\text{kW}$$

$$\text{bsfc} = \dfrac{\text{Fuel consumption}\,(h/hr)}{\text{Power output}}$$

$$= \dfrac{0.35 \times 10^{-3} \times 3600 \times 1000}{9.6}$$

$$\text{bsfc} = 131.25 \text{ g}/\text{kWh}$$

$$= \dfrac{9.6}{0.35 \times 10^{-3} \times 44000} = 100$$

$$\eta_{bth} = 62.3377$$

Steam and Steam Turbines

2.1 Carnot Cycle

The Carnot cycle has a low mean effective pressure because of its really low work output. Thus, one of the varied forms of the cycle to produce the higher mean effective pressure whilst theoretically achieving full Carnot cycle efficiency is the Stirling cycle.

It consists of two isothermal and two constant volume processes. The heat rejection and the addition take place at the constant temperature. The P-V and T-S diagrams for the Stirling cycle are shown in the below figure.

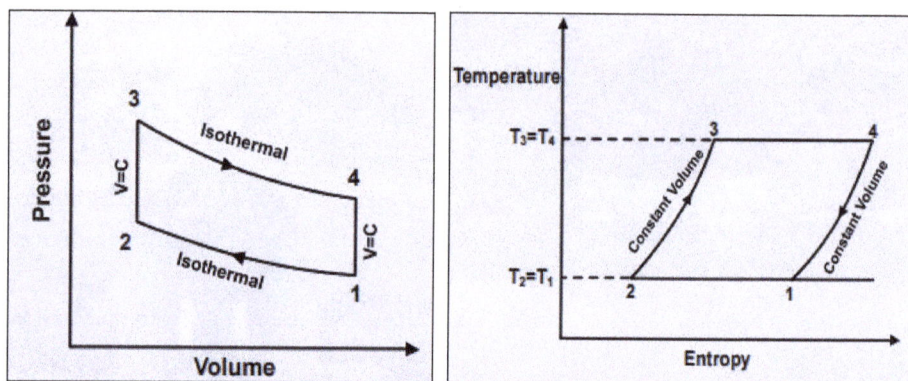

Stirling cycle processes on p-v and T-s diagrams.

Stirling Cycle Processes

The air is compressed isothermally from state 1 to 2 (T_L to T_H):

- The air at state-2 is passed into the regenerator from the top at a temperature T_1. The air passing through the regenerator matrix gets heated from T_L to T_H.

- The air at state-3 expands isothermally in the cylinder until it reaches state-4.

- The air coming out of the engine at temperature enters into regenerator from the bottom and gets cooled while passing through the regenerator matrix at constant volume and it comes out at a temperature T_L, at condition 1 and the cycle is being repeated again.

- It may be shown that the heat absorbed by the air from regenerator matrix during the process 2-3 is equal to the heat given by air to the regenerator matrix during the process 4-1, then exchange of heat with the external source will be only during an isothermal processes.

Now, Net work done = $W = Q_s - Q_R$.

Heat supplied = Q_s = heat supplied during the isothermal process 3-4.

$$= P_3 V_3 \ln\left(\frac{V_4}{V_3}\right); \ r = \frac{V_4}{V_3} = CR$$

$$= mRT_H \ln(r)$$

Heat rejected = Q_R = The heat rejected during isothermal compression process, 1-2,

$$= P_1 V_1 \ln\left(\frac{V_1}{V_2}\right)$$

$$= mR\, T_L \ln(r)$$

$$W_{net} = mR\ln(r)\left[T_H - T_L\right]$$

Now,

$$\eta_{th} = \frac{net}{Q_s} = \frac{mR\ln(r)(T_H \quad T_L)}{m\,R\,\ln(r)\,T_H} \frac{T_H \quad T_L}{T_H}$$

And,

$$\eta_{th} = 1 - \frac{T_L}{T_H}$$

Hence, the efficiency of Stirling cycle is equal to that of Carnot cycle efficiency when both are working with same temperature limits. It is impossible to obtain 100% efficient regenerator and therefore, there will be always 10 to 20 % loss of heat in the regenerator, that decreases the cycle efficiency. Considering regenerator efficiency, the efficiency of the cycle may be written as,

$$\eta_{th} = \frac{R\ln(r)(T_H - T_L)}{R\,T_H\ln(r) + (1 - \eta_R)\,C_V(T_H - T_L)}$$

Where, η_R is known as the regenerator efficiency.

Dual Cycle

This cycle is also known as the dual cycle, which is shown in the figure. Here the heat addition occurs partly at a constant volume and partly at a constant pressure. This cycle is a closer approximation to the behavior of actual Otto and Diesel engines because in the actual engines, the combustion processes do not occur exactly at the constant volume or at constant pressure but rather as in the dual cycle.

Process 1-2: is the Reversible adiabatic compression.

Process 2-3: is the Constant volume heat addition.

Process 3-4: is the Constant pressure heat addition.

Process 4-5: is the Reversible adiabatic expansion.

Process 5-1: is the Constant volume heat rejection.

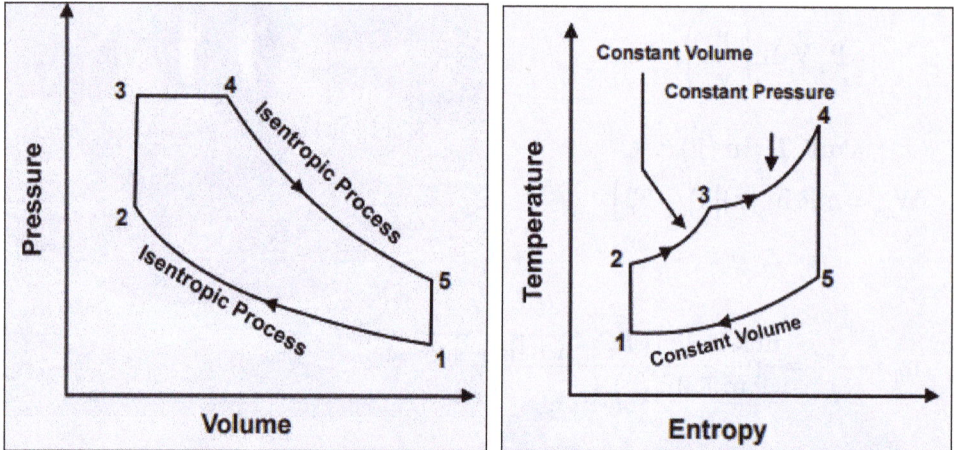

Dual cycle on p-v and T-s diagrams.

The Air Standard Efficiency

Heat supplied $= m\,C_v\left(T_3 - T_2\right) + m\,C_p\left(T_4 - T_3\right)$

Heat rejected $= m\,C_v\left(T_5 - T_1\right)$

Net work done $= m\,C_v\left(T_3 - T_2\right) + m\,C_p\left(T_4 - T_3\right) - m\,C_v\left(T_5 - T_1\right)$

$$\eta_{th} = \frac{m\,C_v\left(T_3 - T_2\right) + m\,C_p\left(T_4 - T_3\right) - m\,C_v\left(T_5 - T_1\right)}{m\,C_v\left(T_3 - T_2\right) + m\,C_p\left(T_4 - T_3\right)}$$

$$\eta_{th} = 1 - \frac{T_5 - T_1}{\left(T_3 - T_2\right) + \gamma\left(T_4 - T_3\right)}$$

Let,

$$\frac{P_3}{P_2}=r_p \; ; \; \frac{V_4}{V_3}=r_c \; ; \; \frac{V_1}{V_2}=r$$

$$T_2 = T_1 \, r^{\gamma-1}$$

$$T_3 = T_2 \, r_p = T_1 \, r^{\gamma-1} \, r_p$$

$$T_4 = T_3 \, r_c = T_1 \, r^{\gamma-1} \, r_p \, r_c$$

$$\frac{T_5}{T_4} = \left(\frac{V_4}{V_5}\right)^{\gamma-1} = \left(\frac{V_4}{V_2} \cdot \frac{V_2}{V_5}\right)^{\gamma-1} = \left(\frac{r_c}{r}\right)^{\gamma-1}$$

$$T_5 = T_4 \left(\frac{r_c}{r}\right)^{\gamma-1} = T_1 \, r_p \, r_c^{\gamma}$$

$$\eta_{th} = 1 - \frac{T_1 \, r_p \, r_c^{\gamma} - T_1}{\left\{\left(T_1 \, r^{\gamma-1} \, r_p - T_1 \, r^{\gamma-1}\right) + \gamma \left(T_1 \, r^{\gamma-1} \, r_p \, r_c - T_1 \, r^{\gamma-1} \, r_p\right)\right\}}$$

$$= 1 - \frac{\left(r_p \, r_c^{\gamma} - 1\right)}{\left\{\left(r_p \, r^{\gamma-1} - r^{\gamma-1}\right) + \gamma \left(r_p \, r_c \, r^{\gamma-1} - r_p \, r^{\gamma-1}\right)\right\}}$$

$$\eta_{th} = 1 - \frac{1}{r^{\gamma-1}} \left\{\frac{r_p \, r_c^{\gamma} - 1}{\left(r_p - 1\right) + \gamma r_p \left(r_c - 1\right)}\right\}$$

From the equation above, it is observed that, a value of $r_p > 1$ results in increased efficiency for a given value of r_c and γ. Hence, the efficiency of the dual cycle lies between the Otto cycle and the Diesel cycle that has the same compression ratio.

H-S Diagrams

The Mollier diagram is plot of enthalpy (h) versus entropy (s) as shown in the following figure. It is also called as the h-s diagram. This diagram consist of a series of constant temperature lines, constant quality lines, constant pressure lines, and constant volume lines. The Mollier diagram is used only when the quality is greater than 50% and for superheated steam. For any state, at least the two properties must be known to determine the other unknown properties of the steam at that state.

The available Mollier diagram is truncated from a point beyond the particular critical point i.e. This shows only a portion of the diagram. In such truncated diagram property of liquid cannot be read.

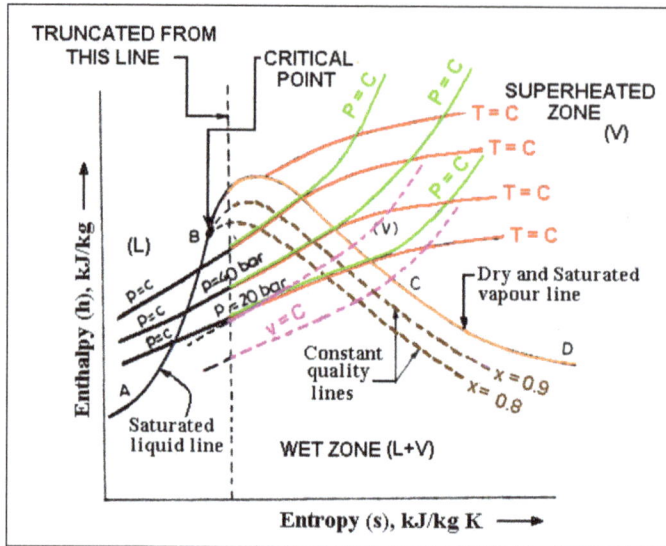

Mollier or Enthalpy-Entropy (h-s) diagram.

2.1.1 Analysis of Various Thermodynamic Processes Under Gone by Steam

When system undergoes the change from one thermodynamic state to final state due change in properties such as temperature, pressure, volume etc., the system is considered to have undergone thermodynamic process. Several types of thermodynamic processes are as follows. Such as isothermal process, adiabatic process, isochoric process, isobaric process and reversible process. These have been described below:

Isothermal Process

When the system undergoes the change from one state to the other, but if its temperature remains constant, the system is said to have undergone isothermal process. For instance, consider our example of hot water in thermos flask, if we remove certain quantity of water from the flask, but keep its temperature constant at 50 degree Celsius, the process is known as the isothermal process.

One other example of an isothermal process is latent heat of vaporization of water. When we heat water to 100 degree Celsius, it will not start boiling instantly. It will keep absorbing heat at a constant temperature; and This heat is called the latent heat of vaporization. Only after absorbing this heat, water at constant temperature, will get converted into steam.

Adiabatic Process

The process in which heat content of the system or certain quantity of the matter remains constant, is known as the adiabatic process. Hence, in an adiabatic process no transfer of heat between the system and its surroundings take place. The wall of the system that does not allow the flow of heat through it, is known as the adiabatic wall, while the wall which allows the flow of heat is known as the diathermic wall.

Isochoric Process

The process in which the volume of the system remains constant, is known as the isochoric process. Heating of gas in a closed cylinder is an example of the isochoric process.

Isobaric Process

The process in which the pressure of the system remains constant is known as the isobaric process. For ex: Suppose there is fuel in piston and cylinder arrangement. When this fuel is burnt the pressure of the gases is generated inside the engine and since more fuel burns more pressure is created. But when the gases are allowed to expand by allowing the piston to move outside, the pressure of the system may be kept constant.

The constant pressure and volume processes are really important. The Otto and diesel cycle are used in the petrol and the diesel engine respectively. They have constant volume and constant pressure processes. In practical situations ideal constant pressure and constant pressure processes may not be achieved.

Reversible Process

In simple words the process which can be reversed back completely is known as a reversible process. This means the final properties of the system should be perfectly reversed back to original properties. The process may be perfectly reversible only if the changes in the process are infinitesimally small. In practical situations it is impossible to trace these extremely small changes in extremely small time, thus the reversible process is also an ideal process. The changes that occur during the reversible process are in equilibrium with each other.

2.2 Vapor Power Cycles

Process 1-2: is the Reversible adiabatic compression process from P_1 to P_2.

Process 2-3: is the Reversible isothermal heat addition process at constant temperature T_H.

Process 3-4: is the Reversible adiabatic expansion process from P_3 to P_4.

Process 4-1: is the Reversible isothermal heat rejection process at constant temperature T_L

The saturated vapor leaves the boiler at state 3, enters the turbine and expands to state 4.

The fluid now enters the condenser, where it is cooled to state 1 and then it is compressed to state 2 in the pump. The efficiency of the cycle is shown as follows:

$$\eta_{carnot} = \frac{T_H - T_L}{T_H} = \left[1 - \frac{T_L}{T_H} \right]$$

Practically, it is very difficult to add or reject heat to or from the working fluid at constant temperature. But, it is comparatively easy to add or reject heat to or from the working fluid at constant pressure.

Therefore, Carnot cycle cannot be used as an idealized cycle for the steam power plants. But, ideal cycle for steam power plant is Rankine cycle in which the heat addition and rejection takes place at the constant pressure process.

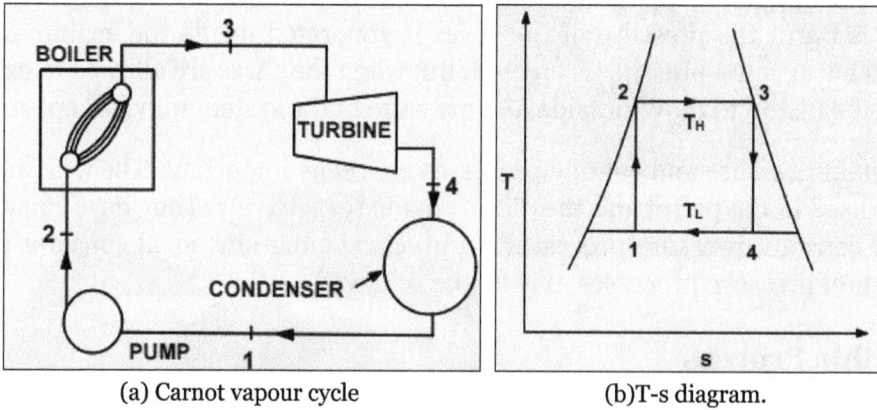

(a) Carnot vapour cycle   (b)T-s diagram.

Comparison of Carnot and the Rankine Cycles

Carnot cycle is a theoretical cycle, that describes a heat engine. Before explaining the Carnot cycle, few terms need to be defined. Heat source is defined as a constant temperature device, that will help in providing infinite heat.

The heat sink is a constant temperature device, which will absorb infinite amount of heat without changing the temperature. The engine is a device or a process, that converts heat from the heat source to work. The Carnot cycle has four steps.

Reversible Isothermal Expansion of the Gas

The engine is thermally connected with the source. The expanding gas absorbs heat from the source and does work on the surroundings in this step. The temperature of gas remains constant.

Reversible Adiabatic Expansion of the Gas

When the system is said to be adiabatic it means that no heat transfer is possible. The engine is taken out of the source and insulated. In this step, the gas does not absorb any heat from the source. The piston continues to do the work on the surrounding.

Reversible Isothermal Compression

The engine is placed on the sink and made to contact thermally. The gas is compressed so that the surrounding is doing a work on the system.

Reversible Adiabatic Compression

The engine is taken out of the sink and then insulated. The surrounding still continues to do work on the system.

In the Carnot cycle, the total work done is given by difference between the work done on the surroundings and the work done by the surroundings. Carnot cycle is the most efficient heat engine in theory. The efficiency of the Carnot cycle depends only on temperatures of the source and sink.

Rankine Cycles

Rankine cycle is a cycle, that converts heat into work. The Rankine cycle is practically used for systems that consist of a vapor turbine. There are four main processes in the Rankine cycle:

- The working of fluid into high pressure from a low pressure.

- The heating of the high pressure fluid into a vapor.

- The vapor expands through a turbine turning the turbine, and thereby generating power.

- The vapor is cooled back inside the condenser.

2.2.1 Thermodynamic Variables Effecting Efficiency and Output of Rankine Cycle

Steam power plants are responsible for production of most electric power in the world, and even small increase in thermal efficiency will result in large savings from the fuel requirements. Hence, every effort is made to improve the efficiency of the cycle on which steam power plant operates. The general idea behind all he modifications to increase the thermal efficiency of the power cycle is the same. They are the average fluid temperature must be as high as possible during heat addition and as low as possible during heat rejection. Next we discuss three ways of accomplishing this for the simple ideal Rankine cycle.

Effect of Reducing the Condenser Pressure

The steam is entered into the condenser as a saturated mixture of vapor and moisture at the turbine back pressure p_2. If this pressure is lowered, the saturated temperature of exhaust steam is decreased, and hence, the amount of heat rejection in the condenser is also decreased. The efficiency of the Rankin cycle is increased by lowering the condenser pressure.

As turbine back pressure p2 decreases to p2' the heat rejection decreases by area 2-3-3'-2'-2. The heat transfer to steam is increased by an area a'-4'-4-a-a'. Thus, the net work done and efficiency of the cycle increases. However, lowering condenser pressure is not without the side effects such as:

- Lowering the back pressure cause an increase in the moisture content of the steam leaving the turbine. It is an unfavorable factor, because, when the moisture content of steam in low pressure stages of the turbine exceeds 10%, there is a decrease in turbine efficiency and erosion of turbine blade may be a serious problem.

- To maintain high vacuum, the air extraction pump will run continuously and its work input will increase, thus reducing the useful work.

Effect of lowering condenser pressure on the Rankin cycle.

Effect of Superheating

Superheating of steam increases mean temperature of heat addition. The effect of superheated steam on the performance of the Rankin cycle is shown. The increase in superheat is shown by the line 1-1'. The hatched area 1-1'-2'-2-1 represents an increase in net work done during the cycle. The area under curve 1-1' represents increase in the heat input. Thus, both the net work done and heat transfer increase as a result of superheating the steam to higher temperature. Therefore, superheating begets higher cycle efficiency.

It is obtained that the specific steam consumption is decreased when steam is superheated. Superheating of steam to higher temperature is desirable, as the moisture content of steam leaving the turbine decreases which is indicated by the state 2'. However, the metallurgical considerations restrict the superheating of steam to a very high temperature.

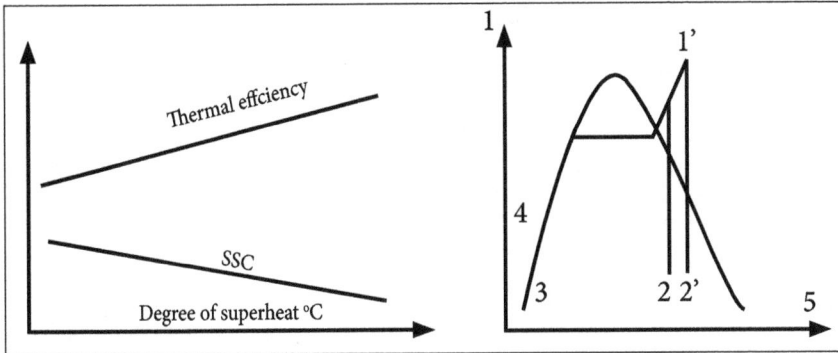

Effect of superheating on ranking cycle.

Effect of Increase in Boiler Pressure

By increasing the boiler pressure, the mean temperature of the heat addition increases and hence raises the thermal efficiency of the cycle. By keeping the maximum temperature T_{max} and the condenser pressure p_2 constant, its boiler pressure increases, the heat rejection decreases by area b'-2'-2-b-b'. The net work done by the cycle remains almost same, therefore, the Rankine cycle efficiency increases, with an increase in maximum pressure. Actually, the operating conditions of the condenser remains unchanged and there is no drastic gain in the work output. Specific steam consumption decreases initially and then increases after reaching a minimum level at 160 bar.

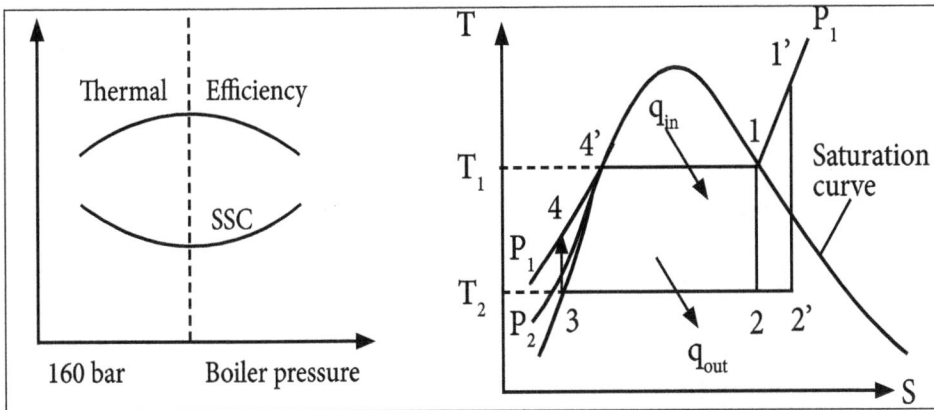

Effective boiler pressure on ranking cycle, performance.

Table: Effect of the operating variables on work done and thermal efficiency of Rankin cycle:

Operating variable	Work done	Efficiency
Decrease in condenser pressure	Increase	Increase
Increase in boiler pressure	No effect	Increase
Superheating steam	Increase	Increase

2.2.2 Analysis of Simple Rankine Cycle and Re-Heat Cycle

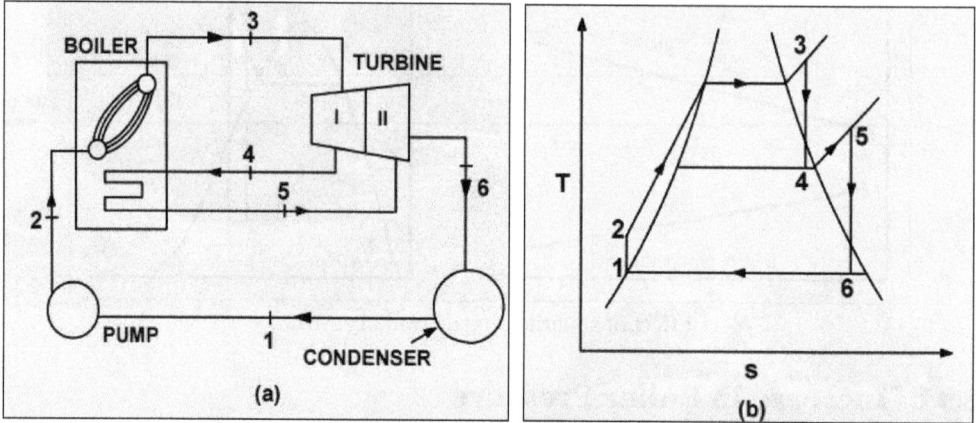

Rankine cycle with reheat.

In reheat Rankine cycle, expansion of steam is carried out in various stages and the steam is reheated by adding heat between stages of turbine. Thus excessive moisture in the low-pressure stages of the turbine is avoided.

The above figure shows the schematic and corresponding T-s, p-v diagrams of a reheat Rankine cycle with two turbine stages. Steam is then expanded from the boiler pressure P_3 to some intermediate pressure P_4 in first stage of the turbine. It is then reheated in the boiler from state 4 to state 5 and then finally expanded from $P_4 = P_5$ to the exhaust pressure $P_1 = P_6$, in the second stage of the turbine. Note that we can employ any number of turbine stages.

The process of reheating does not result in any appreciable gain in the thermal efficiency, because the average temperature of heat addition is not changed. The main advantage is that the moisture content of the steam is reduced to a safe value.

Thermal Efficiency of Reheat Cycle

$$\eta_{reheat} = \frac{\text{Net workdone}}{\text{Heat supplied}}$$

$$= \frac{(h_3 - h_4) + (h_5 - h_6) - (h_2 - h_1)}{(h_3 - h_2) + (h_5 - h_4)}$$

Neglecting pump work,

$$\eta_{reheat} = \frac{(h_3 - h_4) + (h_5 - h_6)}{(h_3 - h_1) + (h_5 - h_4)}$$

2.3 Steam Turbines

Steam Power Plant

A thermal power station is a power plant in which the prime mover is steam driven. Water is heated, and then turns into steam, spins the steam turbine which in turn drives the electrical generator. After it passes through the turbine, the steam is condensed in a condenser and recycled to where it was heated; this is called as the Rankine cycle.

The greatest variation in designing the thermal power stations is because of the different fuel sources. Some power station prefer to use the term energy center because such facilities convert forms of heat energy into electricity. Some thermal power plants also deliver heat energy for the industrial purposes, for district heating or for desalination of water as well as delivering the electrical power.

A large proportion of CO_2 is produced by the world's fossil fired thermal power plants; efforts to reduce these outputs are more and widespread.

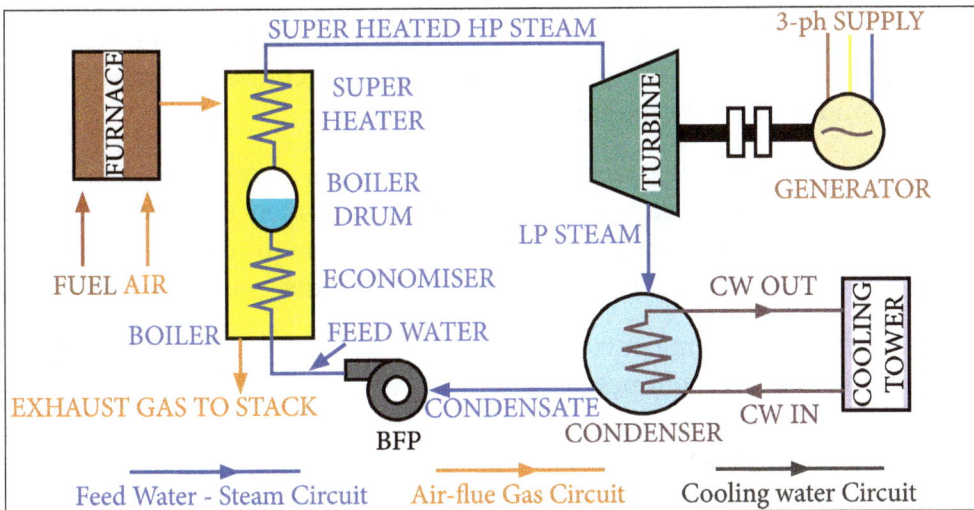

Stream Power plant.

The four main circuits one would come across in any thermal power plant layout are:

- Air and Gas Circuit.

- Coal and Ash Circuit.

- Cooling Water Circuit.

- Feed Water and Steam Circuit.

Air and Gas Circuit

Air from the atmosphere is directed into furnace through the air preheated by the action of a forced draught fan or an induced draught fan. The dust from the air is removed even before it enters the combustion chamber of the thermal power plant layout. The exhaust gases from the combustion heat the air, which goes through a heat exchanger and finally it is let off to the environment.

Coal and Ash Circuit

Coal and Ash circuit in a thermal power plant layout mainly takes care of feeding boiler with coal from the storage for combustion. The ash that is generated during combustion is collected at the back of the boiler and removed to the ash storage by scrap conveyors. The combustion in Coal and Ash circuit is controlled by means of the speed regulation and the quality of coal that enters the grate and the damper openings.

Cooling Water Circuit

The quantity of cooling water that are needed to cool the steam in a thermal power plant layout is significantly high and thus it is supplied from a natural water source like a lake or a river. After passing through screens that remove particles, it is passed through the condenser where the steam is condensed.

The water is finally discharged back into the water source after cooling. Cooling water circuit may also be a closed system where the cooled water is sent through cooling towers for reuse in the power plant. The cooling water circulation in the condenser of a thermal power plant layout helps in maintaining a low pressure in the condenser throughout.

All these circuits are now integrated to form a thermal power plant layout which can generate electricity to meet our needs.

Feed Water and Steam Circuit

The steam produced in the boiler is supplied to the turbines to generate power. The steam that is expelled by the prime mover in the thermal power plant layout is then condensed in a condenser for re-use in the boiler.

The condensed water is forced through a pump into the feed water heaters where it is heated using the steam from different points in the turbine. To make up for the lost steam and water while passing through the various components of the thermal power plant layout, feed water is supplied through external sources. Feed water is purified in a purifying plant to reduce the dissolve salts that could scale the boiler tubes.

The boiler feed water used in the steam boiler is a means of transferring heat energy from burning fuel to the mechanical energy of the spinning steam turbine.

The total feed water consists of a recirculate condensate water and purified makeup water. The makeup water is highly purified before use, as the metallic materials it contacts are subject to corrosion at high temperatures and pressures.

A system of ion exchange demineralizers and water softeners produces pure water so that it coincidentally becomes an electrical insulator, with conductivity in range of 0.3–1.0 micro Siemens per centimeter.

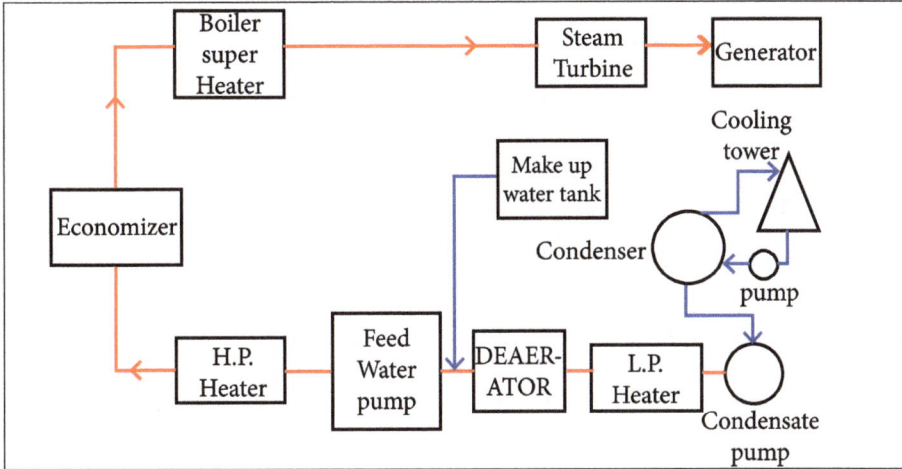

Water steam flow diagram.

2.3.1 Classification of Steam Turbines

Superheated steam is being generated from the steam boiler and then is distributed to steam turbine generator. Steam turbine do have some classifications. The classification of steam turbine is given in the following:

1. Based on the Steam Flow Direction

- Radial steam turbine, the steam turbine which has direction of steam flow perpendicular to the axis of shaft.

- Axial steam turbine, the steam turbine which has steam flow direction parallels to the axis of shaft.

2. Based on the Working Principal

Impulse steam turbine, the steam turbine which the rotation of its blades is caused by steam where the velocity of the steam had been increased by the nozzle. The impulse steam turbine includes:

- Curtis steam turbine.

- Parson steam turbine.

- De-Laval steam turbine.

- Zoelly/Rateau steam turbine.

Reaction steam turbine, a steam turbine in which the rotation of its blades is caused by the reaction of its blades due to the steam flow itself.

3. Based on the Exit Steam

- Direct condensation steam turbine.

- Non condensing steam turbine with direct flow.

- Back pressure steam turbine.

- d. Condensation extraction steam turbine.

- e. Back pressure extraction steam turbine.

4. Based on Steam Pressure

- Middle pressure steam turbine, the turbine with pressure up to 40 ata.

- Low pressure steam turbine, the turbine with pressure up to 2 ata.

- Super critical pressure steam turbine, the turbine with presses exceeds 225 ata.

- High pressure steam turbine, the turbine with pressure 40 – 170 ata.

- Very high steam turbine, the turbine with pressure exceeds 170 ata.

2.4   Impulse Turbine and Reaction Turbine

The steam jets are being directed at the turbine, bucket shaped rotor blades where the pressure exerted by the jets cause the rotor to rotate and then causes the velocity of the steam to reduce as it imparts its kinetic energy to the blades.

The blades change the direction of flow of steam, but its pressure remain constant as it passes through the rotor blades. The cross section of chamber between the blades remains to be constant. Impulse turbines are hence also known as the constant pressure turbines. The next series of fixed blades reverses the direction of the steam before it passes to second row.

The rotor blades of the reaction turbine are shaped and arranged such that the cross section of the chambers formed between the fixed blades diminishes from the inlet side towards exhaust side of blades. The chambers between rotor blades essentially forms nozzles so that when the steam progresses through the chambers its velocity increases ,at the same time its pressure decreases, just because of the nozzles formed by the fixed blades.

Therefore, the pressure decreases in both the fixed and the moving blades. The steam in a jet from between the rotor blades, creates a reaction force on the blades which in turn creates the turning moment on turbine rotor, just like in steam engine.

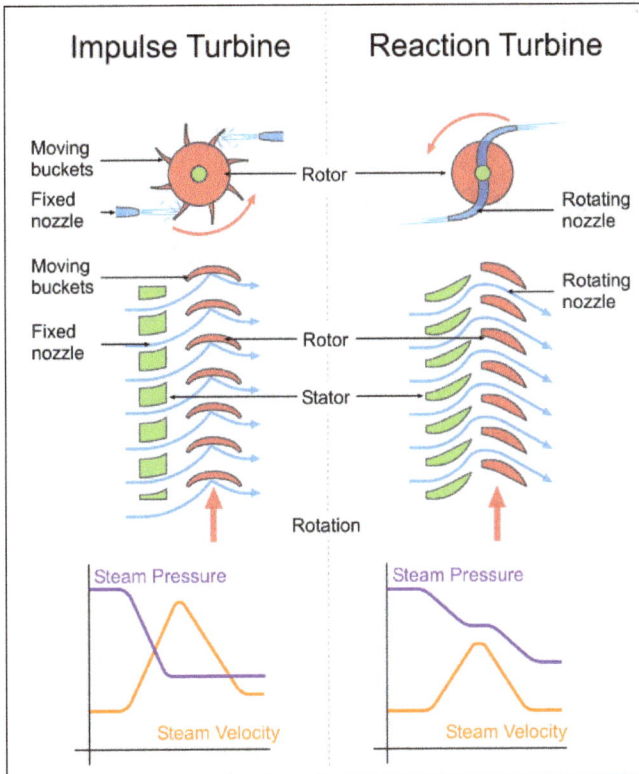

Impulse Turbine and Reaction Turbine.

Differences Between Impulse and Reaction Turbines

S. No	Impulse Turbine	Reaction Turbine
1.	In Impulse Turbine all hydraulic energy is converted into kinetic energy by a nozzle and it is is the jet so produced which strikes the runner blades.	In Reaction Turbine only some amount of the available energy is converted into kinetic energy before the fluid enters the runner.
2.	The velocity of jet which changes, the pressure throughout remaining atmosphere.	Both pressure and velocity changes as fluid passes through a runner. Pressure at inlet is much higher than at outlet.
3.	Impulse Turbine operates at high water heads.	Reaction turbine operate at low and medium heads.
4.	Impulse Turbine have more hydraulic efficiency.	Reaction Turbine have relatively less efficiency.
5.	Needs low discharge of water.	Needs medium and high discharge of water.
6.	Water flow is tangential direction to the turbine wheel.	Water flows in radial and axial direction to turbine wheel.

7.	The turbine is always installed above the tail race and there is no draft tube used.	Reaction turbines are generally connected to the tail race through a draft tube which is a gradually expanding passage. It may be installed below or above the tail race.
8.	Impulse turbine involves less maintenance work.	Reaction turbine involves more maintenance work.
9.	Degree of reaction is zero.	Degree of reaction is more than zero and less than or equal to one.

Energy Losses in Steam Turbines

Enumerate the energy losses in steam turbines.

The energy losses in steam turbines are:

- Losses in regulating values.
- Losses due to steam friction.
- Losses due to mechanical friction.
- Losses due to leakage.
- Losses due to wetness of steam.
- Losses due to radiation.
- Residual velocity losses.
- Carry over losses.

Degree of Reaction

The degree of reaction turbine state is defined as the ratio of heat drop over moving blades to the total heat drop in the stage.

$$R_d = \frac{\text{Heat drop in moving blades}}{\text{Heat drop in the stage}}$$

Problems

1. Steam issues from the nozzles of a De Laval turbine with a velocity of 1000 m/sec. The nozzle angle is 20°. Mean blade velocity is 400 m/sec. The blades are symmetrical. The mass flow rate is 1000 kg/h. Friction factor is 0.8, nozzle = 0.95. Let us determine:

- Blade angles.
- Axial thrust on the rotor turbine.
- Work done per kg of steam.

- Power developed.
- Blade efficiency.
- Stage efficiency.

Solution:

Given:

$$V_1 = 1000 \text{ m/s}$$

$$a = 20°$$

$$V_0 = 4000 \text{ m/sec}$$

$$m = 1000 \text{ kg/hr} = 0.27 \text{ kg/sec}$$

$$\eta_{nozzle} = 0.95$$

$$k = \frac{V_{r2}}{V_{r1}} = 0.8$$

To Find:

- Blade angle $(\theta,)$.
- Axial thrust (F_y).
- Work done per kg of steam (W_D).
- Power developed (P).
- Stage efficiency $\left(\eta_{stage}\right)$.
- Blade efficiency $\left(\eta_{blade}\right)$.

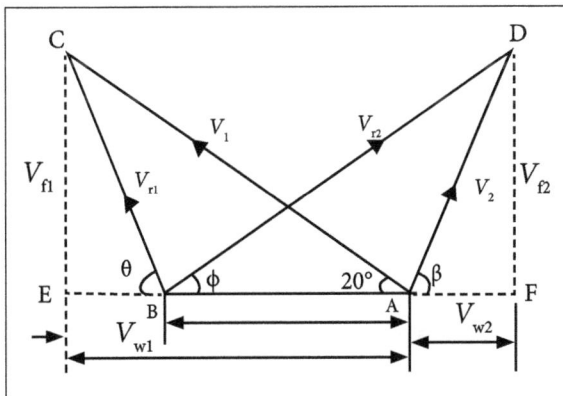

From ΔBCE,

$$V_{\omega 1} = V_1 \cos 20° = 1000 \cos 20° = 939.69 \text{ m / s}$$

$$V_{f1} = V_1 \sin 20° = 1000 \sin 20° = 342.02 \text{ m / s}$$

From ΔACE,

$$V_{r1} = \sqrt{V_{f1}^2 + \left(V_{\omega 1} - V_b\right)^2} = \sqrt{342.02^2 + \left(939.69 - 400\right)^2}$$

$$V_{r1} = 638.93 \text{ m/s}$$

$$\tan \theta = \frac{V_{f1}}{V_{\omega 1} - V_b} = \frac{342.02}{939.63 - 400}$$

$$\theta = 32.21$$

But,

$$\frac{V_{r2}}{V_{r1}} = K$$

$$V_{r2} = 0.8 \times 638.93$$

$$= 511.14 \text{ m/s}$$

For symmetrical blading,

$$\theta \, \alpha = 32° \times 21$$

From ΔADF,

$$V_{f2} = V_{r2} \sin f$$

$$= 511.14 \times \sin 32°.21'$$

$$= 273.50 \text{ m / s}$$

Similarly,

$$V_b + V_{\omega 2} = V_{r2} \cos 32°21$$
$$400 + V_{\omega 2} = 511.14 \cos 32°21$$
$$V_{\omega 2} = 31.80 \text{ m/s}$$

From ΔBDF,

$$V_2 = \sqrt{V_{f2}^2 + V_{\omega 2}^2}$$

$$= \sqrt{273.50 + 31.80^2}$$

$$= 275.34\, m/s$$

Driving force,

$$F_x = m\left(V_{\omega 1} + V_{\omega 2}\right)$$

$$= 0.27\left(939.69 + 31.80\right)$$

$$= 262.30\, N$$

Axial thrust,

$$F_y = m\left(V_{f1} - V_{f2}\right)$$

$$= 0.27\left(342.02 - 273.50\right)$$

$$F_y = 18.5\, N$$

Power $P = m V_b \left(V_{\omega 1} + V_{\omega 2}\right)$

$$= 262.30 \times 400$$

$$= 104.92\, kW$$

Work done/kg of steam,

$$W_D = V_b\left(W_{\omega 1} + V_{\omega 2}\right)$$

$$= 400\left(939.69 + 31.80\right)$$

$$W_D = 388.59\, kJ$$

Blade Efficiency

$$\eta_b = \frac{m\left(V_{\omega 1} + V_{\omega 2}\right) \times V_b}{\dfrac{m V_1^2}{2}}$$

$$= \frac{104.92 \times 1000}{\dfrac{0.27 \times 1000^2}{2}}$$

$$= 77.71\, \%$$

Stage efficiency,

$$\eta_{stage} = \eta_b \times \eta_v$$
$$= 0.7771 \times 0.95$$
$$= 73.82\ \%$$

Result,

$$\theta\ \alpha = 32°21$$
$$F_y = 18.5\ N$$
$$W_D = 388.59\ kJ$$
$$P = 104.92\ kW$$
$$\eta_{blade} = 77.71\%$$
$$\eta_{stage} = 73.82\%.$$

2. A One stage of an impulse turbine consists of a converging nozzle ring and one ring of moving blades. The nozzles are inclined at 22° to the blades whose tip angles are both 35°. If the velocity of steam at exit from the nozzle is 660 m/s, let us calculate the blade speed so that the steam passes without shock. Let us also find the diagram efficiency neglecting losses if the blades are run at this speed.

Solution:

Given data:

$$\alpha = f = 22°$$
$$q = \beta = 35°$$
$$V_a = 660\ m / s$$

To find:

- Blade speed.
- Diagram efficiency.

From the figure,

$$V_b = \text{blade velocity (Scale: 1cm = 60 m/s)}$$
$$= 4.2 \times 60$$
$$= 252\ m / s$$

Blade efficiency or diagram efficiency,

$$\eta_b = \frac{2\,V_D\left(V_{\omega 1} + V_{\omega 2}\right)}{V_1^2}$$

$$V_{\omega 1} = 10.2 \times 60 = 612\,\text{m/s}$$

$$V_{\omega 2} = 1.7 \times 60 = 102\,\text{m/s}$$

$$\eta_b = \frac{2 \times \left(612 + 102\right) \times 252}{660^2}$$

$$\eta_b = 82.6\%$$

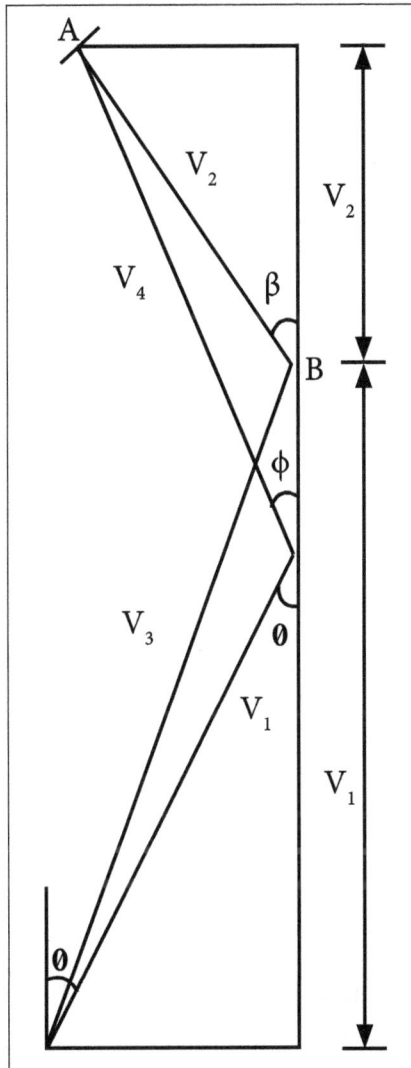

3. A 300 kg/min of steam (2 bar, 0.98 dry) flows through a given stage of a reaction turbine. The exit angle of fixed blades as well as moving blades is 20° and 3.68 kW of power is developed. It the rotor speed is 360 rpm and tip leakage is 5 percent, let us

calculate the mean drum diameter and the blade height. The axial flow velocity is 0.8 times the blade velocity.

Solution:

Given data:

$$m_1 = 300 \text{ kg}/\text{min} = 5 \text{ kg}/\text{sec}$$
$$N = 360 \text{ rpm}$$
$$P = 2 \text{ bar}$$
$$x = 0.98$$
$$V_f = 6.8 \text{ Vb}$$
$$P = 3.68 \text{ kW} = 3.68 \times 10^3 \text{ W}$$

To find:

- Drum diameter d.

- Blade height, h.

Since the tip leakage is 5 percent, therefore actual mass of steam flowing over the blades,

$$m = 5 - (5 \times 0.05)$$
$$= 4.75 \text{ kg}/S$$

We know that,

$$V_b = \frac{\pi d_m N}{60}$$
$$= \frac{\pi d_m \times 360}{60}$$
$$= 18.849 \, d_m \text{ m/s}$$
$$V_f = 0.8 \, V_b$$
$$= 0.8 \times 18.849 \, d_m$$
$$V_f = 15.0796 \, d_m \text{ m/s}$$

Let us draw the combined velocity triangle,

1. First of all, draw a horizontal line and cutoff AB equal to 18.849 dm to some suitable scale representing the blade velocity Vb.

2. Now draw inlet velocity triangle ABC on the base AB with,

$$\alpha = 20° \text{ and BC} = \frac{V_f}{\sin 20°}$$

$$BC = \frac{V_f}{\sin 20°} = \frac{15.0796\, d_m}{0.342}$$

BC = 44.08 d_m to the scale.

3. Similarly, draw outlet velocity triangle on the same base AB with<doubt> ø = 20° and,

$$V_{r_1} = \frac{V_{f_1}}{\sin 20°} = \frac{15.0796\, d_m}{0.342} = 44.08\, d_m \text{ to the scale}$$

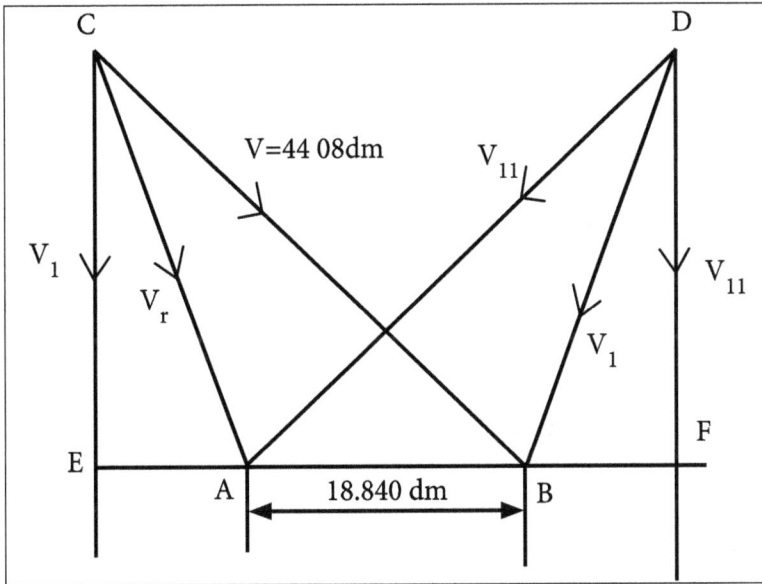

4. Form C and D draw perpendicular to meet the line AB produced at E and F.

By measurement from velocity triangle, we find that change in the velocity of whirl,

$$(V_w + V_{w1}) = 57\, d_m$$

We know that power developed (P),

$$p = m(V_w + V_{w1})V_b$$

$$3.68 \times 10^3 = 4.75 \times 57\, cd_m \times 18.849\, d_m$$

$$3.68 \times 10^3 = 5100.93\, d_m^2$$

$$d_{m2} = 1.3861$$

$$d_m = 1.177\, m$$

$$V_{f_1} = V_f = 15.0796 \, d_m$$

$$= 15.0796 \times 1.177$$

$$V_f = 17.753 \text{ m}/\text{s}$$

From steam tables, corresponding to a pressure of 2 bar, we find that specific volume of steam,

$$V_g = 0.885 \text{ m3}/\text{kg}$$

Mass of steam flow m,

$$m = \frac{\pi \, dm \, V_{f_1}}{x \, V_g}$$

$$4.75 = \frac{\pi \times 1.177 \times h \times 17.753}{0.98 \times 0.885}$$

$$h = 0.06257 \text{ m}$$
$$h = 627 \text{ mm}$$

Height of the blades h = 62.7 mm

Drum diameter,

$$d = d_m - h$$
$$= 1.177 - 0.0627$$
$$d = 1.114 \text{ m}$$

4. A Steam enters the blade row of an impulse turbine with a velocity of 600 m/s at an angle of 25° to the plane of rotation of the blades the mean blade speed is 250 m/s. The blade Angle at the exit side is 30°. The blade friction loss is 10%. Let us determine (i) The blade angle inlet, (ii) The work done per kg of steam (iii) blade efficiency.

Solution:

Given data:

Steam velocity = 600 m/s

Plane of rotation of the blades =25°

Mean blade speed = 250 m/s

Exit blade angle = 30°

To Find:

- Blade angle inlet.

- The work done per kg of steam.

- Blade efficiency.

The velocity diagram is constructed as described below. AB is drawn to scale to represent the blade velocity.

$U = 250$ m/s. AC's drawn at 25° to AB to represent steam velocity. BC. Now represents the relative velocity V_{r_1}. The angle CBE is measured as 41° and this is the blade exit angle:

- BF is marked as 0.9 BC to obtain the relative velocity at exist. BD is drawn at 30° to BA representing the blade angle at exit. Joining AD and drawing perpendicular to AB from D complete the velocity triangles.

GE is measured and converted to velocity. This length represents the charge in the velocity of whirl.

This is found as 585 m/s.

- The work done/kg,

$$= U V_w = 250 \times 585$$

$$= 146.25 \frac{kW}{kg}$$

- Blade efficiency,

$$= \frac{2 U V_w}{V_{a_1}^2} = \frac{2 \times 250 \times 585}{600^2}$$

$$= 81.25\%$$

2.5   Compounding in Turbines

Compounding of steam turbines is the method in which energy from the steam is extracted in a number of stages rather than single stage in a turbine. A compounded steam turbine has multiple stages i.e. it has more than one set of nozzles and rotors, in series, keyed to the shaft or fixed to the casing, so that either the steam pressure or jet velocity is absorbed by the turbine in number of stages.

"The method in which energy from steam is extracted in more than single stage is known as Compounding. A multi-stage turbine i.e. which has more than one set of rotors and nozzles is known as the compounded turbine."

Velocity Compounding of the Impulse Turbine

The velocity compounded impulse turbine has both moving and fixed blades. The moving blades are keyed to turbine shaft and the fixed blades are fitted to the Casing. The high pressure steam from the boiler is expanded in nozzle where the pressure energy is converted into kinetic energy.

The high velocity steam is directed on the first set of moving blades and as steam flows over the blade, it imparts some of its momentum to blades and then loses some velocity. Some part of high K.E is absorbed by the blades and there is no change in velocity of steam as it passes through the fixed blades. The steam then goes to next set of moving blades and the same process is repeated until all the energy of steam is absorbed. The figure below shows the velocity compounding of impulse turbine.

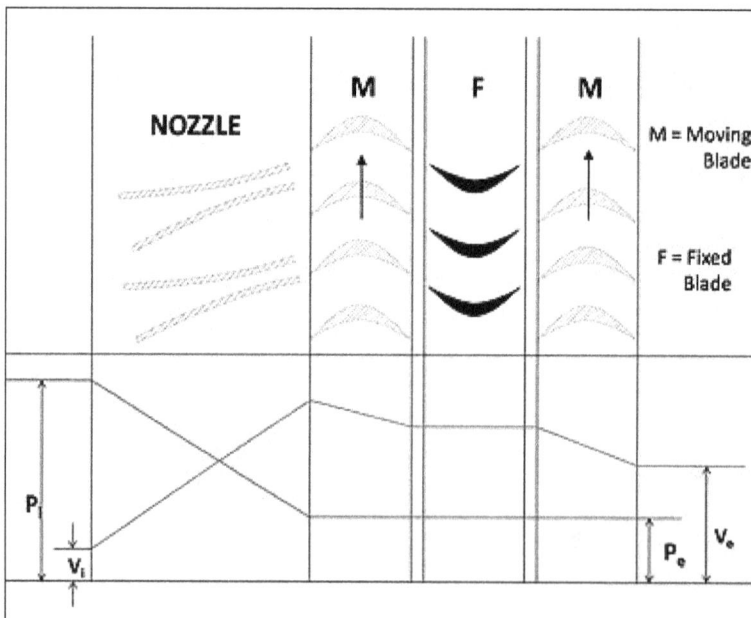

Pi and Po are pressure of steam at inlet and outlet and Vi and Vo are velocities of steam at inlet.

Pressure Compounding of Impulse Turbine

This is used to solve the problem of high blade velocity in the single-stage impulse turbine. It consists of alternate rings of nozzles and turbine blades. The nozzles are fitted to casing and then the blades are keyed to the turbine shaft. In this type of compounding, steam is expanded more than once.

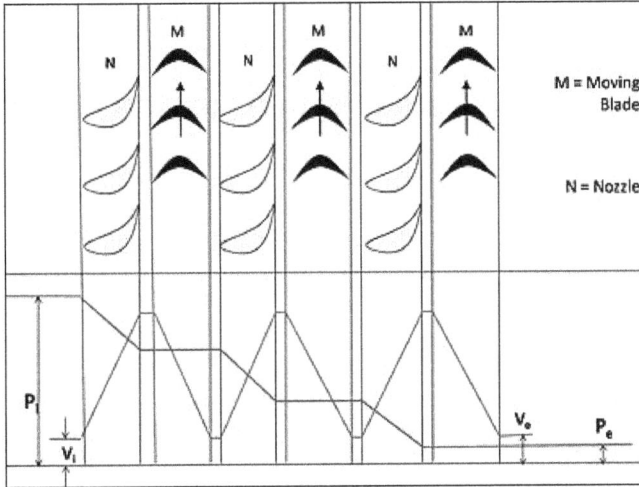

Now, here the high pressure steam is fed to nozzle where it is partially expanded i.e. pressure is decreased and the velocity is increased .When this steam is passed over the set of blades, where almost all its velocity is absorbed and pressure remains constant during this period . This process is repeated until the condenser pressure is achieved.

2.5.1 Velocity Diagrams for Simple Impulse and Reaction Turbines

Pressure Compounding

When four simple impulse turbines are connected in series, the total enthalpy drop is divided equally among the stages. So, the pressure drop only occurs in the nozzle whereas there is no pressure drop in blades.

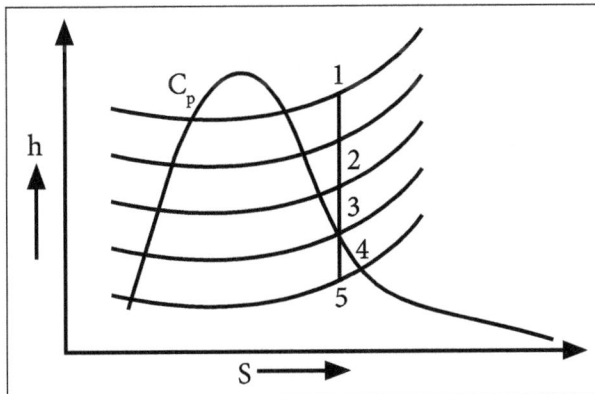

Enthalpy drop in each stage will be equal,

$$\therefore \ h_1 - h_2 = h_2 - h_3 = h_3 - h_4 = h_4 - h_5$$

So,

$$h_1 - h_2 = \frac{h_1 - h_5}{4}$$

$\therefore$ The velocity of steam at exit from the first row of nozzle is given by,

$$V_1 = \sqrt{2000(h_1 - h_2)}$$

Where h_1 and h_2 in kJ/kg,

$$= \sqrt{2000\left(\frac{h_1 - h_5}{4}\right)}$$

$$= \frac{1}{2}\sqrt{2000(h_1 - h_5)} \qquad \qquad \ldots(1)$$

But for a single stage turbine, the velocity of steam at exit of the nozzle,

$$V_1 = \sqrt{2000(h_1 - h_5)} \qquad \qquad \ldots(2)$$

From equation (1) and (2) we can infer that the velocity of steam leaving the nozzles in each stage of 4-stage turbine is half of that for a single stage turbine. For 9-stage turbine, it will be one-third.

So, each impulse turbine operating at its maximum blading efficiency is given by,

$$\frac{V_b}{V_1} = \frac{\cos x}{2}$$

For n-stages, the enthalpy drop per stage will be,

$$(\Delta h)_{stage} = \frac{(\Delta h)_{total}}{n} = \frac{h_1 - h_n}{n}$$

or,

$$\text{Number of stages} = \frac{(\Delta h)_{total}}{(\Delta h)_{stage}}$$

Velocity Compounding

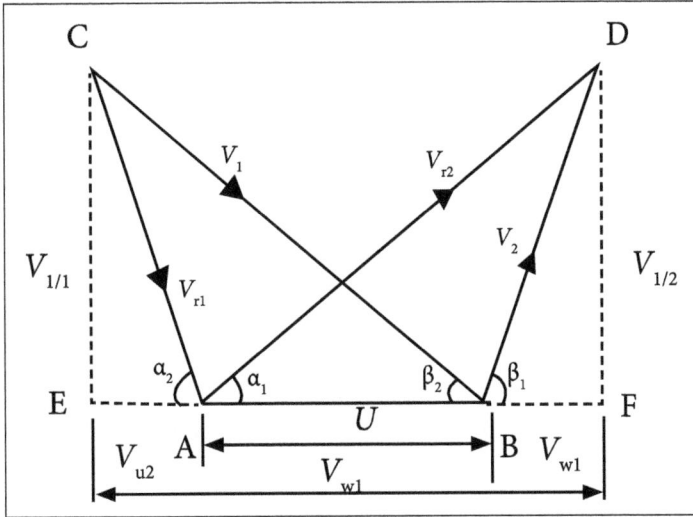

Velocity diagram for the first row of moving blades.

The kinetic energy of steam jets ($1/2$ m V_{12}) at nozzle exit is partially converted into work in the first row of moving blades with velocity differences from V_1 to V_2. Again kinetic energy $\left(\dfrac{1}{2}mV_3^2\right)$ of the exiting steam from the first row of moving blades is converted into work is the next row of moving blades and so on.

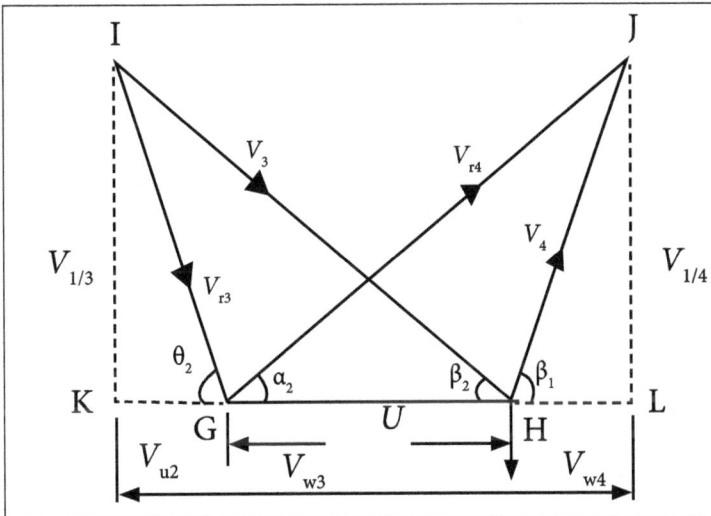

Velocity diagram for second row of moving blades.

Already we know that,

$$\frac{V_{r_2}}{V_{r_1}} = k$$

From this diagram,

$$\text{Workdone}_I = m\left(V_{w_2} + V_{w_2}\right)V_b$$

Axial thrust,

$$F_{Y_1} = m\left(V_{f_1} - V_{f_2}\right)$$

Kinetic energy of steam supplied for the first stage,

$$\text{K.E}_I = \frac{1}{2}mV_1^2$$

The same friction factor is considered for the next row of moving blades.

$$\therefore \frac{V_3}{V_2} = k \text{ and also } \frac{V_{r_4}}{V_{r_3}} = k$$

$$\text{Work done}_{II} = m\left(V_{w_3} + V_{w_4}\right)V_b$$

Axial thrust,

$$F_{YII} = m\left(V_{f_3} - V_{f_4}\right)$$

Kinetic energy of stem supplied for second stage,

$$\text{K.E}_{II} = \frac{1}{2}mV_S^2$$

Total efficiency of steam turbine,

$$\eta = \frac{\text{Workdone}_I + \text{Workdone}_{II}}{\text{K.E}_I + \text{K.E}_{II}}$$

$$\therefore \eta = \frac{2V_b\left(V_{w_1} + V_{w_2} + V_{w_3} + V_{w_4}\right)}{V_1^2 + V_3^2}$$

Similarly, total axial thrust,

$$F_Y = F_{Y_I} + F_{Y_{II}}$$

$$= m\left(V_{f_1} + V_{f_2} + V_{f_3} + V_{f_4}\right)$$

$$= m\left[\left(V_{f_1} + V_{f_3}\right) - \left(V_{f_2} + V_{f_4}\right)\right]$$

Gas Turbines

3.1   Gas Turbines: Simple Gas Turbine Plant

The gas turbine is used to make mechanical energy from a combustible fuel. In the gas turbines, the mechanical energy comes in the form of a rotating shaft. This shaft has an enormous amount of power and torque.

The combustion (gas) turbines being installed in many of today's natural-gas-fueled power plants are complex machines, but basically involve three main sections:

The compressor draws air into the engine, pressurizes it and then feed it to the combustion chamber at speeds of hundreds of miles per hour.

The combustion system, typically made up of rings of fuel injectors that inject a steady stream of fuel into the combustion chambers where it mixes with air. The mixture is then burned at a temperatures of more than 2000 degrees F. The combustion produces high temperature, high pressure gas stream that enters and expands them through the turbine section.

The turbine is an intricate array of alternate stationary and rotating aero foil-section blades. As hot combustion gas expands through the turbine, it spins the rotating blades. The rotating blades perform a dual fuvvnction, first is that, they drive the compressor to draw more pressurized air into the combustion section and secondly they spin a generator to produce electricity.

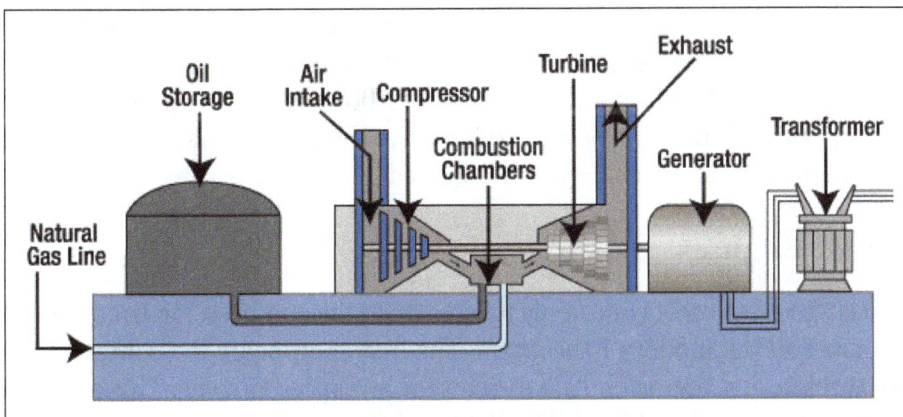

Combustion turbine power plant.

Land based gas turbines are of two types:

- Heavy frame engines.
- Aero-derivative engines.

Heavy frame engines are characterized by lower pressure ratios and tend to be physically large. Pressure ratio is the ratio of compressor discharge pressure and inlet air pressure. Aero-derivative engines are derived from the jet engines, just like the name implies and they are operated at very high compression ratios. Aero-derivative engines tends to be very compact and it is useful where smaller power outputs are needed.

Since the large frame turbines have higher power outputs, they can produce a larger amounts of emissions and should be designed to achieve low emissions of the pollutants, like Nitrogen Oxide. One key to turbine's fuel to power efficiency is that, the temperature at which it operates. Higher temperatures usually mean higher efficiencies, which in turn, can lead to more economical operation.

Gas flowing through a typical power plant turbine may be as hot as 2300 degrees F, but some of those critical metals in the turbine will withstand temperatures only as hot as 1500 to 1700 degrees F. Therefore, air from the compressor may be used for cooling the key turbine components, and thereby reducing the ultimate thermal efficiency.

nA simple cycle gas turbine achieves energy conversion efficiencies that ranges between 20 and 35 percent. With higher temperatures achieved in the Department of Energy's turbine program, future hydrogen and syngas fired gas turbine combined cycle plants are more likely to achieve efficiencies of 60 percent or more. When waste heat is captured from those systems for heating or for industrial purposes, the overall energy cycle efficiency may approach 80 percent.

Compressed Air Energy Storage

One modern development seeks to improve efficiency in another way, by separating the compressor and the turbine with a compressed air store. In conventional turbine, up to half the generated power is used driving the compressor. In a compressed air energy storage configuration, power from a wind farm or bought on the open market at a time of low demand and low price, is being used to drive the compressor and the compressed air released to operate the turbine when required.

Turbo Shaft Engines

Turbo shaft engines are most commonly used to drive the compression trains and are used to power almost all the modern helicopters. The first shaft bears the compressor and the high speed turbine, and the second shaft bears the low speed turbine. This arrangement is used to increase the speed and flexibility of its power output.

Radial Gas Turbines

In 1963, Jan Mowill initiated the development at Kongsberg Vapenfabrikk in Norway. Severasl successors have made a good progress in the refinement by means of this mechanism. Owing to a configuration that keeps heat away from certain types of bearings, the durability of the machine is improved while the radial turbine is well matched in the speed requirement.

Scale Jet Engines

Scale jet engines are the scaled down versions of the early full scale engine. It is also called as the miniature gas turbines or the micro-jets.

Advantages

As they are smaller than coal or nuclear plants, gas power plants may be built faster and at a lower cost. Gas turbine systems also require comparitively less water than steam power plants and they can be easily converted into combined cycle power plants, which are even more efficient.

There are various advantages in using a gas power plant to generate the electrical power as compared to other systems. Gas turbine power plants can be started up and run at full capacity in only 10 to 20 minutes, making them well suited as a backup plants to be used in the utility companies which requires more additional electricity immediately.

Disadvantages

- The operating temperature in gas turbines is higher than other power plant systems and may shorten the lifespan of some of the system components.

- Gas turbine power plants have many disadvantages. The power that is required to drive the compressor reduces the net outputs, consuming more fuel to do the same and equal amount of work.

- Moreover, as the thermal energy is wasted when the exhaust is released, the efficiency levels of gas turbine plants are lower than those of other types of the power plants.

3.1.1 Ideal Cycle: Closed and Open Cycle

By analyzing the ideal gas turbine cycles provides a clear understanding of parameters affecting the power output and the thermal efficiency, and these idealized calculations are readily extended to deal with the real cycles.

The following assumptions are made:

- Compression and expansion processes are reversible and also they are adiabatic.

- The change in the kinetic energy of the working fluid between the inlet and the outlet of each component is negligible.

- There are no pressure losses in ducting or in the combustion changer.

- The working fluid consist of the same composition throughout the cycle and this working fluid is a perfect gas with constant specific heats; air maybe assumed to be the fluid.

- The mass flow of gas is constant throughout the cycle.

Gas Turbine are Mainly Divided Into Two Group

1. Constant Pressure Combustion Gas Turbine

- Open cycle.

- Closed cycle.

2. Constant Volume Combustion Gas Turbine

The most of the sector open cycle gas turbine plants are used. Closed cycle plants were introduced at one stage because of their ability to burn low cost fuel.

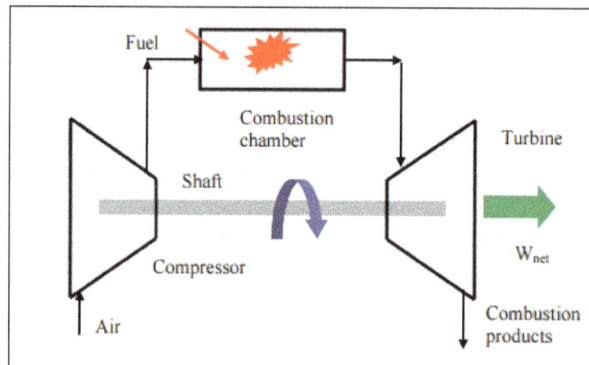

Open Gas-Turbine Cycle.

A schematic diagram of an easy gas rotary engine power plant is shown in the above figure. Air is drawn from atmosphere into a mechanical device, wherever it is compressed reversibly and adiabatically. The comparatively high pressure is then used in burning the fuel in combustion chamber.

The air fuel quantitative relation is quite high to limit the temperature of the burnt gases getting into the rotary engine. The gases then expands isentropically in the rotary engine. A portion of the work obtained from the rotary engine is used to drive the mechanical device and hence the auxiliary drive and rest of the ability output is that the internet power of the turbine plant.

A gas rotary engine plant works with the help a Brayton or joule cycle. This cycle was originated by joule, a British engineer for use in a hot air reciprocating engine and later in about 1870 an American engineer George Brayton tried this cycle in a gas turbine. This cycle has two constant pressures and two adiabatic processes. The P-V and T-S diagrams of the cycle are as shown in figure.

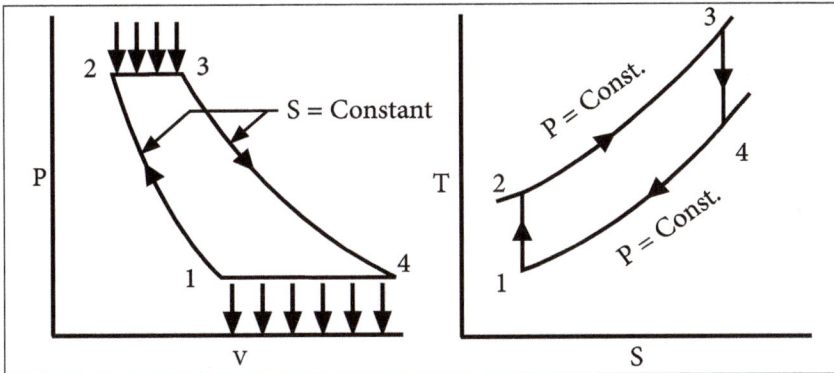

P-v and T-s diagram of air stander gas-turbine cycle.

- Process 1 – 2: Isentropic compression in the compressor

- Process 2 – 3: Constant pressure heat addition in the combustion chamber

- Process 3 – 4: Isentropic expansion in the turbine

- Process 4 -1: Constant pressure heat rejection in the atmosphere or cooling of air in the Intercooler (closed cycle).

Close Cycle Gas Turbine Engine

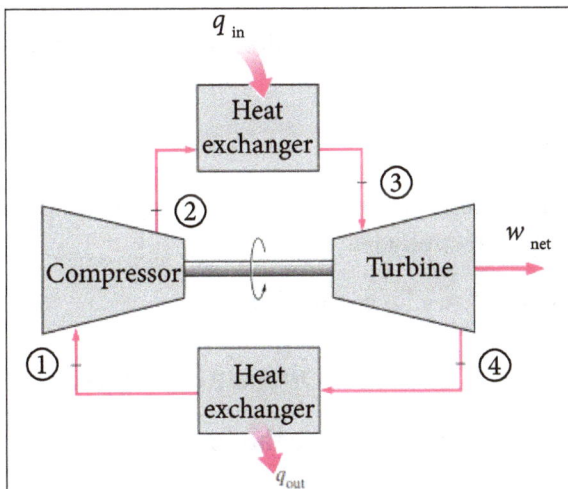

Close cycle gas turbine Engine.

Because of its confined, fixed amount of gas, the closed cycle gas turbine is not an internal combustion engine. In the closed cycle system, combustion cannot be sustained and the normal combustor is replaced with a second heat exchanger to heat the compressed air before it enters into the turbine.

The heat is supplied by an external source such as a nuclear reactor, the fluidized bed of a coal combustion process, or some other heat source. Closed cycle systems using gas turbines have been proposed for missions to Mars and other long term space applications.

Advantages of Closed Cycle

- Higher thermal efficiency.

- Reduced size.

- No contamination.

- Improved heat transmission.

- Improved part load h.

- Lesser fluid friction.

- No loss of working medium.

- Large output.

- Less fuel.

Disadvantages of Closed Cycle

- Quality.

- Large amount of cooling water is needed. This limits its use of stationary installation or marine use.

- Dependent system.

- The weight of the system pre kW developed is high comparatively, therefore not economical for moving vehicles.

- Requires the use of a very large air heater.

3.1.2 Efficiency, Work Ratio and Optimum Pressure Ratio

The ideal cycle for the simple gas turbine is the Brayton cycle, i.e. cycle 1234 in Figure given below,

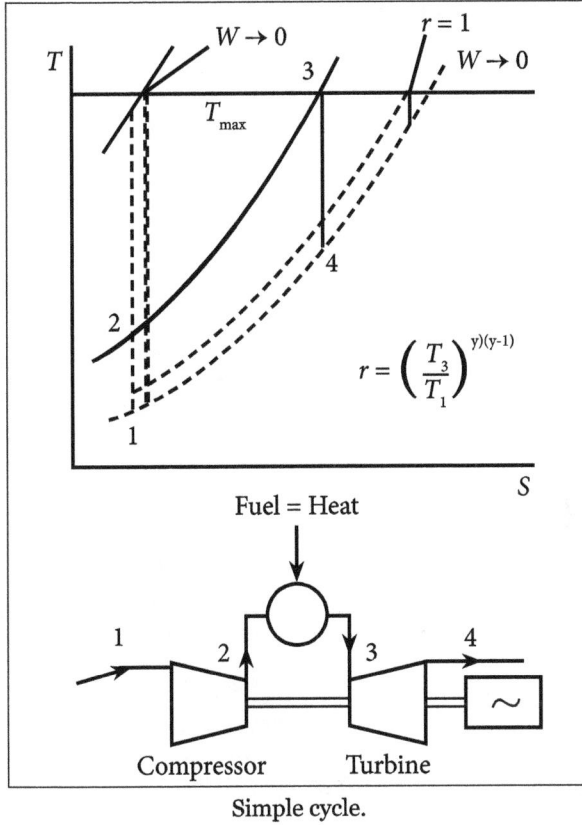

Simple cycle.

The relevant steady flow equation is,

$$Q = (h_2 - h_1) + \tfrac{1}{2}(C_2^2 - C_1^2) + W$$

Where Q and W are the heat and work transfers per unit mass flow, h is the specific enthalpy and C the fluid velocity. Applying this to each component, we have,

Compressor work input $= h_2 - h_1 = C_p(T_2 - T_1)$

Combustor heat input $= h_3 - h_2 = C_p(T_3 - T_2)$

Turbine work output $= h_3 - h_4 = C_p(T_3 - T_4)$

The cycle efficiency is,

$$\eta = \frac{\text{net work output}}{\text{heat supplied}} = \frac{c_p(T_3 - T_4) - c_p(T_2 - T_1)}{c_p(T_3 - T_2)}$$

Making use of the isentropic p-T relation we have,

$$T_2 / T_1 = r^{(\gamma-1)/\gamma} = T_3 / T_4$$

Where r is pressure ration $P_2/P_1 = P_3/P_4$ the cycle efficiency is then readily shown is given by,

$$\eta = 1 - \left(\frac{1}{r}\right)^{(\gamma-1)/\gamma}$$

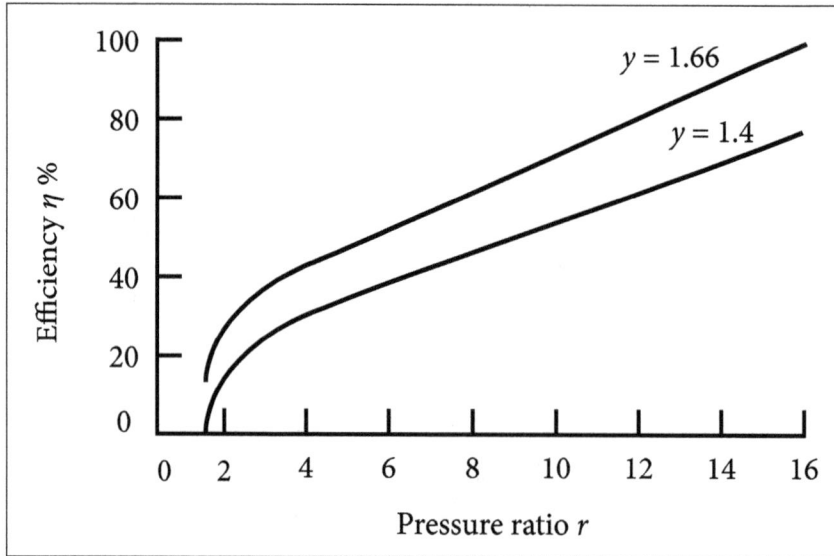

Efficiency for simple cycle.

The efficiency thus depends only on the pressure ratio and the nature of the gas; Figure shows above the result for $\gamma = 1.40$, the value normally assumed for air.

The specific work output W, upon which the size of the plant for a given power depends, is found to be a function of both pressure ratio and the maximum cycle temperature T_3.

Thus,

$$W = c_p\left(T_3 - T_4\right) - c_p\left(T_2 - T_1\right)$$

Which can be expressed as,

$$\frac{W}{c_p T_1} = t\left(1 - \frac{1}{r^{(\gamma-1/\gamma)}}\right) - \left(r^{(\gamma-1/\gamma)} - 1\right)$$

Where t= T_3/T_2 : T_1 is normally atmospheric temperature. It is therefore convenient to plot the specific work output in non-dimensional form (W/Cp T,) as a function of r and as in Figure given below. The value of T_3 and hence t, that can be used in practice, is dependent on the maximum temperature which the highly stressed turbine blades can stand for the required working life: it is often known as the 'metallurgical limit'. Early gas turbines used values of t around 4, but the use of air-cooled turbine blades allowed to be increased to between 5 and 6.

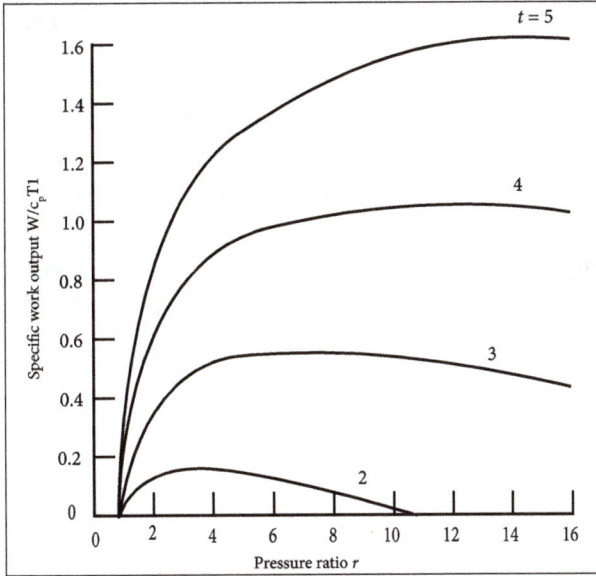

Specific work output for simple cycle.

3.2 Actual Cycle

Irreversibility exists in the actual cycle. The most important differences are pressure drop in combustion chamber, deviations of actual compressor and turbine from idealized isentropic compression/expansion, and the pressure drop in combustion chamber,

$$\eta_c = \frac{w_s}{w_a} \cong \frac{h_{2s} - h_1}{h_{2a} - h_1}$$

$$\eta_T = \frac{w_a}{w_s} \cong \frac{h_3 - h_{4a}}{h_3 - h_{4s}}$$

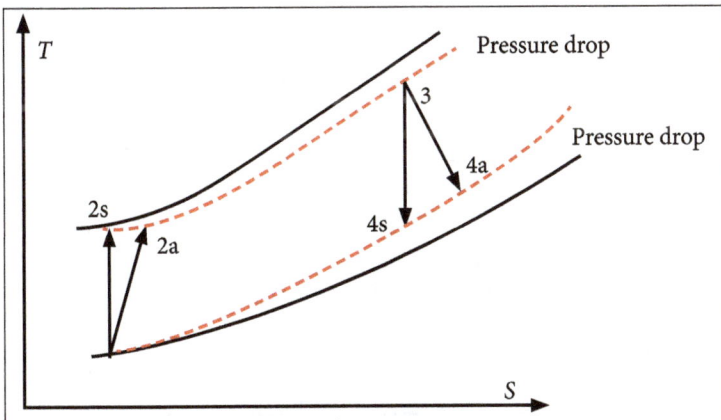

Actual Brayton cycle.

3.2.1 Analysis of Simple Cycles and Cycles with Inter Cooling, Reheating and Regeneration

The network output of the cycle can be increased by reducing the work input to the compressor and thereby increasing the work output from turbine. Using a multi-stage compression along with intercooling reduces work input the compressor. As the number of stages increases, the compression process becomes almost isothermal at the compressor inlet temperature, and compression work decreases. Same way utilizing multistage expansion with reheat will increase the work produced by turbines.

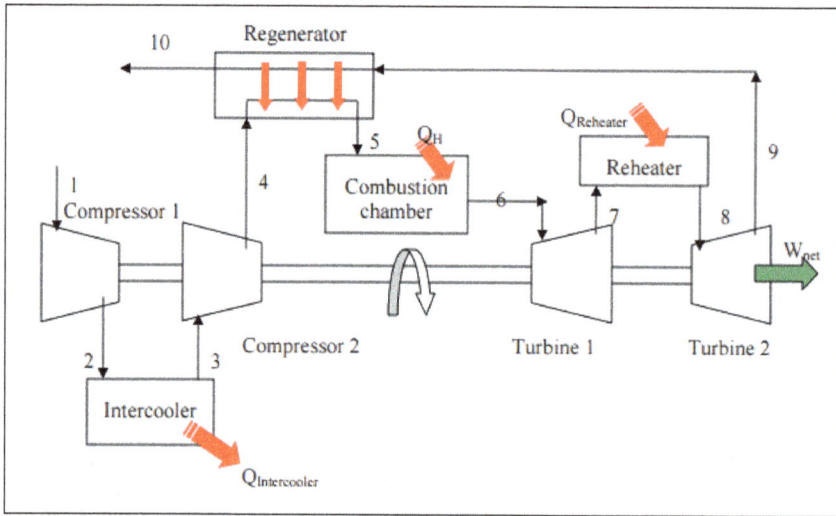

A gas-turbine engine with two-stage compression with intercooling, two-stage expansion with reheating, and regeneration.

If intercooling and reheating are used, regeneration becomes more attractive as a greater potential for regeneration exists. The back work ratio of a gas-turbine is improved as a result of intercooling and reheating. But, intercooling and reheating decreases thermal efficiency unless they are accompanied along with regeneration.

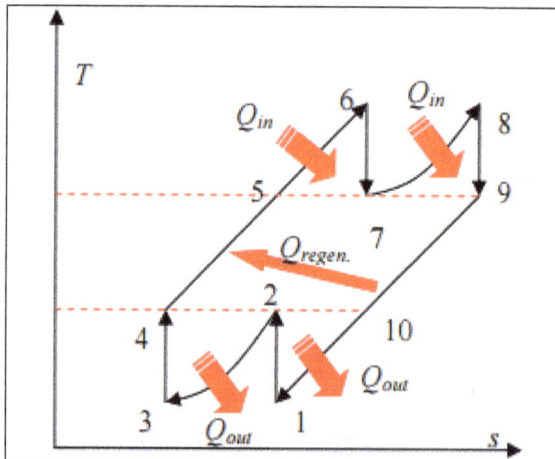

T-s diagram for an ideal gas-turbine cycle with intercooling, reheating, and regeneration.

Reheat Cycle

In reheat cycle, steam is extracted from a suitable point in the turbine and reheated with the help of flue gases in the boiler.

Regenerative Cycle

The feed water is heated with the help of steam in a reversible manner, the temperature of steam and water is same at any section. This type of heating is known as regenerating.

Reheating

Reheating occurs in the turbine and this is a way to increase turbine work without changing the compressor work or melting the materials from which the turbine is constructed. If a gas turbine has a high pressure and a low pressure turbine at the back end of the machine, a reheater is used to "reheat" the flow between the two turbines. This can increase the efficiency by 1-3%. Reheating in a jet engine may be done by adding an afterburner at turbine exhaust, and thereby increasing the thrust at the expense of the greatly increased fuel consumption rate.

Regeneration

Regeneration involves the installation of a heat exchanger through which the turbine exhaust gases pass. The compressed air is then heated in the exhaust gas heat exchanger, before the air flow enters the combustor.

If the regenerator is properly designed, the efficiency will be increased over the simple cycle value. However, the relatively high cost of this regenerator must also be taken into account. Regenerated gas turbines increase efficiency 5-6% and are even more effective in improved part-load applications.

Gas Turbine with Regenerator

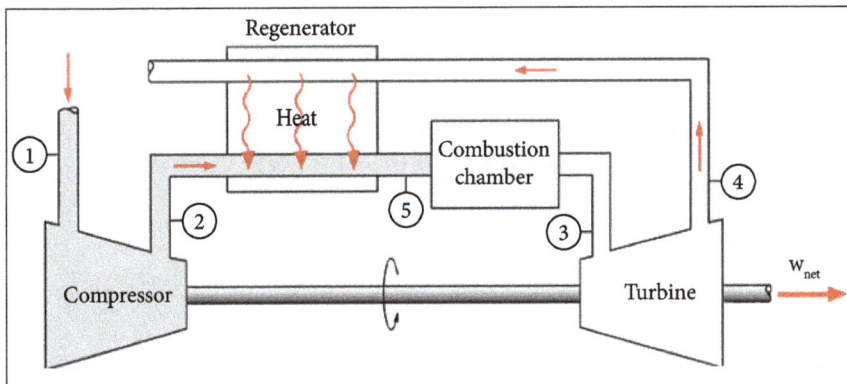

Gas Turbine with Regenerator.

Intercooling

Intercooling also involves the use of a heat exchanger. An intercooler is a heat exchanger which cools the compressor gas during the compression process. For instance, if the compressor consists of a high and a low pressure unit, the intercooler is mounted between them to cool down the flow and decrease the work necessary for the compression in the high pressure compressor. The cooling fluid could be atmospheric air or water (e. g, sea water in the case of a marine gas turbine). It is shown that the output of a gas turbine is increased using a well-designed intercooler.

Effect of Regeneration of a Steam Power Plant

The effects of regeneration of a steam power plant are:

- Improving thermal efficiency.

- Effective utilization of heat.

As shown in figure above. $T_1 = T_3$, $T_2 = T_4$

In an ideal regenerator, $T_5 = T_9$. In practice (actual regenerator), $T_5 < T_9$.

$T_8 = T_6$, $T_7 = T_9$ the network input to a two-stage compressor is minimized when equal pressure ratios are maintained across each stage. That is,

$$\frac{P_2}{P_1} = \frac{P_4}{P_3}$$

This procedure also maximizes the turbine work output,

$$\frac{P_6}{P_7} = \frac{P_8}{P_9}$$

Problems

1. A steam power plant runs on a single regenerative heating process. The steam enters the turbine at 30 bar and 400°C and the steam fraction is withdrawn at 5 bar. The remaining steam exhaust at 0.10 bar to the condenser. Let us calculate the efficiency, steam rate of the power plant. Neglect pump work.

Solution:

$P_1 = 30$ bar

$T_1 = 400°C$

$P_2 = 5$ bar

$P_3 = 0.1$ bar

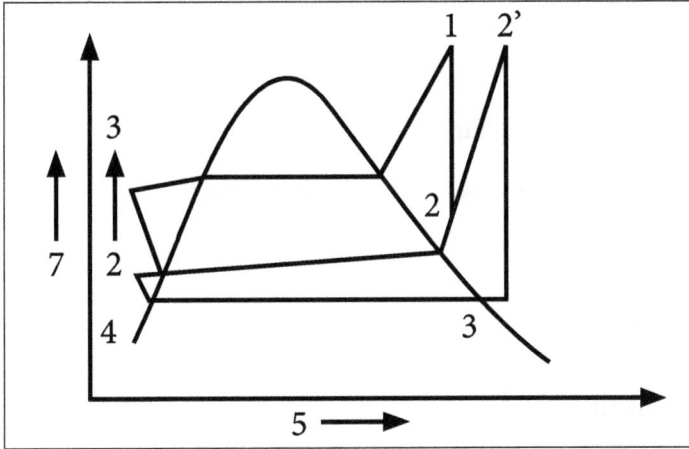

Properties of Steam from Steam Table

At 30 bar and 400°C

$h_1 = 3232.5$ kJ / kg ; $S_P = 6.925$ kJ / kg K

At 5 bar,

$T_{sat} = 151.8°C$

$h_f = 640.1$ kJ / kg ; $h_{fg} = 2107.4$ kJ / kg

$S_f = 1.86$ kJ / kg k ; $S_{fg} = 4.959$ kJ / kg k

$S_g = 6.8191$ kJ / kg

At 0.1 bar,

$T_{sat} = 45.83°C$

$h_f = 191.8$ kJ / kg ; $h_f = 2392.9$ kJ / kg

$S_f = 0.649$ kJ / kg k ; $S_{fg} = 7.402$ kJ / kg k

$S_g = 8.151$ kJ / kg

$1-2 \Rightarrow$ isentropic expansion

$$S_1 = S_2 = 6.925 \text{ kJ / kg k}$$

$$S_s > S_g \text{ at 5 bar}$$

So, steam is at super heat condition.

By linear interpolation the entropy super heat temperature.

At 5 bar is 173°C,

$$\therefore h_2 = 2794.83 \text{ kJ / kg}$$

$3-2 \Rightarrow$ isentropic expansion

$$S_2 = S_3 = 6.925 \text{ kJ / kg}$$

$$S_3 < S_g \text{ at 0.1 bar}$$

$\therefore$ Steam is in wet condition,

$$\therefore S_3 = S_{f_3} + x_3 \times S_{fg_3}$$

$$x_3 = \frac{S_3 - S_{f_3}}{S_{fg_3}} = \frac{6.925 - 0.649}{7.502}$$

$$= 0.84$$

$$\therefore h_3 = h_{f_3} + x_3 + h_{fg_3} = 191.8 + 0.84 \times 2392.9$$

$$= 2201.83 \text{ kJ / kg}$$

$$h_4 = 191.8 \text{ kJ / kg}$$

Mass of steam bed,

$$m = \frac{h_{f_2} - h_4}{h_2 - h_4} = \frac{620.1 - 191.8}{2794.83 - 191.8}$$

$$= 0.1722 \text{ kJ / kg of steam}$$

Work done by the turbine with regeneration,

$$WRg = (h_1 - h_2) + (1 - m)\ (h_2 - h_3)$$

$$= (3232.5 - 2794.83) + (1 - 0.1722)\ \times (2794.83 - 2201.830)$$

$$= 928.55\ kJ/kg$$

Efficiency of cycle with regeneration,

$$\eta_{reg} = \frac{\omega}{h_1 - h_{f_2}} = \left(\frac{9x - 55}{(3232.5 - 640.1)} 200 \right)$$

$$= 35.82\%$$

Steam rate with regeneration,

$$SSC_{reg} = \frac{3600}{N_{reg}} = \frac{3600}{928.55}$$

$$= 3.877\ kJ/k\omega - hr$$

2. Steam enters the turbine at 3 MPa and 400°C and is condensed at 10 KPa. Some quantity of steam leaves the turbine at 0.6 MPa and enters open feed water heater. Let us compute the fraction of the steam extracted per kg of steam and cycle thermal efficiency.

Solution:

Given:

$$P_1 = 3\ MPa\ ;\ T_1 = 400°C\ ;$$

$$P_2 = 0.6\ MPa\ ;\ P_3 = 10\ KPa$$

From superheated steam table,

At $P_1 = 3\ M_{pa}$ and 400°C

$$h_1 = 3232.5\ kJ/kg\ ;\ S_1 = 6.925\ kJ/kg\ k$$

We know that,

$$S_1 = S_2 = 6.925\ kJ/kg$$

But at $P_2 = 6\ bar$,

$$S_g = 6.758\ kJ/kg\ K$$

Now,

$S_2 > S_g$. So, the steam is again in superheated state.

From Mollier diagram, corresponding to 6.925 entropy at 6 bar the enthalpy is found,

$h_2 = 2630 \text{ kJ / kg}$

We know that,

$\Rightarrow S_1 = S_3 = 6.925 \text{ kJ / kgk}$

At $P_3 = 10$ KPa,

$h_{f3} = 191.85 \text{ kJ / kg}$

$h_{fg3} = 2392.9 \text{ kJ / kg}$

$S_{f3} = 0.649 \text{ kJ / kg K}$

$S_{fg3} = 7.502 \text{ kJ / kg k}$

$V_{f3} = 0.00101 \text{ m3 / kg}$

We know that,

$S_1 = S_3 = S_{f3} + x_3 \times S_{fg3}$

$6.925 = 0.649 + x_3 \times 7.502 \quad x_3 = 0.837$

$h_3 = 6_{f3} + x_3 \times h_{fg3}$

$= 191.8 + 0.837 \times 2392.9 = 2194.66 \text{ kJ / kg}$

$h_4 = h_{f3} = 191.8 \text{ kJ / kg}$

Pump work during 4 – 5,

$h_5 = 0.5959 + 191.8$

$h_5 - h_4 = V_{f3} \left[P_2 - P_3 \right] = 192.4 \text{ kJ / kg.}$

$= 0.00101 \times \left[600 - 10 \right] = 0.5959 \text{ kJ / kg}$

Amount of steam bleed,

$m = \dfrac{h_6 - h_5}{h_2 - h_5} = \dfrac{670.4 - 192.4}{2630 - 192.4} = 0.196 \text{ kg / kg of steam}$

At 6 bar, $V_f = 0.00101 \, m^3 / kg$

$$W_{P6} = h_7 - h_6$$

$$= V_{f2} \left[P_1 - P_2 \right] = 0.00101 \left[3000 - 600 \right]$$

$$= 2.424 \, kJ / kg$$

$$h_7 = 2.424 + 561.47 = 672.824 \, kJ / kg$$

Regenerative Rankine Cycle efficiency,

$$\eta_{regenerative} = \frac{\left(h_1 - h_7 \right) - \left(1 - m \right) \left(h_3 - h_{f3} \right)}{\left(h_1 - h_7 \right)}$$

$$= \frac{\left(3232.5 - 2630 \right) - \left(1 - 0.196 \right) \left(192.4 - 191.8 \right)}{\left(3232.5 - 672.824 \right)}$$

$$\eta_{regenerative} = 37.09\%$$

3. In a reheat steam cycle, the maximum steam temperature is limited to 773 K. The condenser pressure is 10 kPa and the quality at turbine exhaust is 0.8778. Had there been no reheat, the exhaust quality would have been 0.7592. Assuming ideal processes, let us determine (i) reheat pressure (ii) the boiler pressure (iii) the cycle efficiency (iv) the steam rate.

Solution:

Given:

$T_1 = 733 \, k$, $P_4 = 10 \, kpa$, $x_4 = 0.8778$ (with reheat)

$x_4 = 0.7592$ (without reheat)

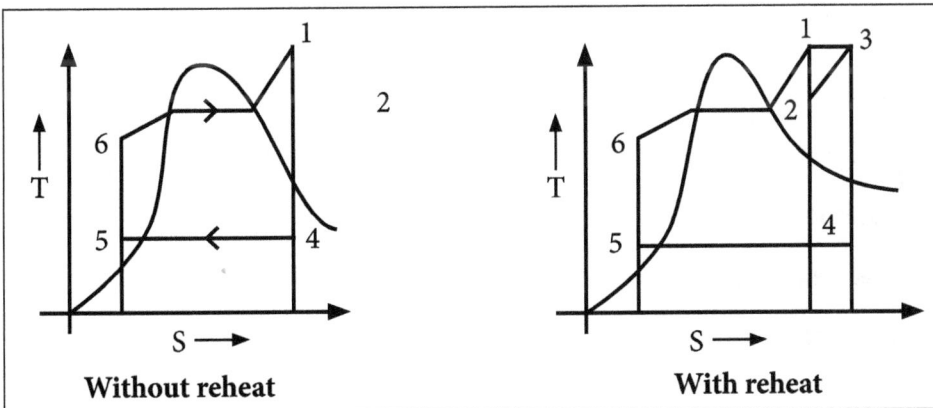

Without reheat   With reheat

From steam table at 1okpa,

$$h_{f_2} = 191.8 \, kJ/kg \qquad h_{fg_2} = 2392.6 \, kJ/kg$$
$$S_{f_2} = 0.649 \, kJ/kg\,k \qquad S_{fg_2} = 7.502 \, kJ/kg\,k$$
$$V_{f_2} = 0.001010 \, m^3/kg$$

We know that,

Case (i) with reheat,

$$S_4 = S_{f4} + x_4 \times S_{fg4}$$

$$S_4 = 0.649 + 0.8778 \times 7.502 = 7.234 \, kJ/kg\,k$$

$$h_4 = h_{f4} + x_4 \times h_{fg4}$$

$$h_4 = 2292.29 \, kJ/kg$$

Case (ii) without reheat,

$$S_4 = S_{f4} + x_4 \times S_{fg4}$$

$$S_4 = 6.35 \, kJ/kg\,k$$

We know that,

$$S_1 = S_4 = 6.35 \, kJ/kg\,k \text{ without reheat.}$$

From Mollier diagram,

By linear interpolation,

Boiler Pressure, P_1 = 140 bar = 14 MPa

$$h_1 = 3330 \, kJ/kg$$

We know that,

$$S_3 = S_4 = 6.35 \, kJ/kg\,k \text{ with reheat from Mollier diagram.}$$

By linear interpolation,

Reheat pressure P_2 = 120 bar = 120 MPa,

$$h_3 = 3470 \, kJ/kg$$

$$h_2 = 2850 \, kJ/kg$$

Pump work,

$$W_P = V_{f_4}\left(P_1 - P_2\right) = 0.0010101 \times \left(14 \times 1000 - 10\right)$$

$$W_P = 14.13 \text{ kJ / kg}$$

Heat supplied,

$$Q_S = h_1 - \left(h_{f_4} + W_P\right) + \left(h_1 - h_2\right)$$

$$Q_S = 3744.07 \text{ kJ / kg}$$

Work done,

$$W = \left(h_1 - h_2\right) + \left(h_3 - h_4\right) - W_p$$

$$= \left(3330 - 2850\right) + \left(3470 - 2292.29\right) - 14.13 = 1643.58 \text{ kJ / kg}$$

Thermal efficiency $= \dfrac{W}{Q_S} = \dfrac{1643.58}{3744.07}$

$$= 44\%$$

Steam rate, SSC $= \dfrac{3600}{1643.58}$

$$= 2.19 \text{ kg / kw hr}$$

4. Consider a steam engine power plant operating on the ideal reheat Rankine cycle. Steam enters the high pressure turbine at 16 MPa and 873 K and is condensed in the condenser at a pressure 10 KPa. If the moisture content of the steam at the exit of the low-Pr turbine is not to exceed 10.4%. Let us determine (i) the pressure at which the steam should be reheated and (ii) the thermal efficiency of the cycle, assuming the steam is reheated to the inlet temperature of the high-pressure turbine.

Solution:

$$P_1 = 16 \text{ MPa} = 160 \text{ bar}$$

$$P_3 = 0.1 \text{ bar}$$

$$T_1 = 600°C \text{ (873 K)}$$

The steam at the exit of low Pr turbine (moisture content) = 10.4%

$$\therefore x_4 = 100 - 10.4 = 89.6\% \text{ (or) } x_4 = 0.896$$

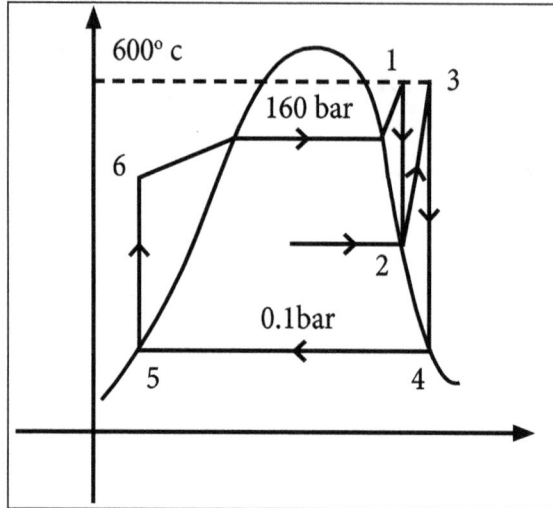

At point 1, 160 bar Pr and 600°C

$$h_1 = 3571 \text{ kJ / kg and } s_1 = 6.639 \text{ kJ / kgk}$$

At point 4, 0.1 bar Pr

$$S_4 = S_{f4} + X_4 \, S_{fg4}$$

$$= 0.649 + \left(0.896 \times 7.502\right)$$

$$\therefore \ s_4 = 7.371 \text{ kJ / kgk}$$

$$h_4 = h_{f4} + X_4 \, h_{fg4}$$

$$= 191.8 + \left(0.896 \times 2392.9\right)$$

$$h_4 = 2335.838 \text{ kJ / kg}$$

$$s_3 = s_4 \left(\because \text{ process } 3-4 \text{ is isentropic}\right)$$

$$\therefore \ s_3 = 7.371 \text{ kJ / kgk}$$

From superheated tables, At $S_3 = 7.371$ kJ/kgk and 600°C

Pressure, $P_2 = 40$ bar

$\therefore$ at point 3, 40 bar and 600°C from tables

$$h_3 = 3672.8 \text{ kJ/kg}$$

$$s_1 = s_2 (\because \text{ process } 1 - 2 \text{ is isentropic})$$

$$\therefore \ s_2 = 6.639 \text{ kJ/kgk}$$

From steam pressure tables,

At 40 bar, s_g = 6.069 kJ/kgk

$\therefore$ steam is in S. H. state

For superheated steam, C_p = 2.1

$$\therefore s_2 = s_{g2} + C_p \, l_n \left[\frac{T_{sup}}{T_{sat}} \right]$$

$$6.639 = 6.069 + (2.1) \, l_n \left[\frac{T_{sup}}{250.3} \right]$$

$$h_2 = h_{g2} + C_p \left(T_{sup} - T_{sat} \right)$$
$$= 2800.3 + (2.1) \, (328.35 - 250.3)$$

$$h_2 = 2964.21 \text{ kJ/kg}$$

$$h_5 = h_{f4}$$

$$h_5 = 191.8 \text{ kJ/kg}$$

$$h_6 = h_{f4} + v_{f4} \left(p_1 - p_2 \right)$$
$$= 91.8 + \left((0.001252) \, (160 - 0.1) \times 100 \right)$$

$$h_6 = 211.819 \text{ kJ/kg}$$

$$W_T = \left(h_1 - h_2 \right) + \left(h_2 - h_4 \right)$$
$$= (3571 - 2964.21) \, (3672.8 - 2335.838)$$

$$W_T = 1943.752 \text{ kJ/kg}$$
$$W_p = 0.001252 \, (160 - 0.1) \times 100$$
$$W_p = 20.01948 \text{ kJ/kg}$$

$$\eta = \frac{W_T - W_P}{\left(h_1 - h_6 \right) + \left(h_3 - h_2 \right)} = \frac{1943.752 - 20.01948}{(3571 - 211.819) + (3672.8 - (2964.21))}$$

$$\eta = 47.29\%$$

Pressure at which the steam should be superheated = P_2 = 40 bar

Thermal efficiency of the cycle, η = 47.29%

4

Impact of Jets and Pumps

4.1 Introduction

Water turbines are widely used throughout the world to generate power. This type of water turbine is referred to as a Pelton wheel, one or more water jets are directed tangentially on to vanes or buckets that are fastened to the rim of the turbine disc.

The impact of the water on the vanes generates a torque on the wheel causing it to rotate and develop power. Although the concept is essentially simple, such turbines can generate considerable output at high efficiency.

Powers in excess of 100 MW and hydraulic efficiency is greater than 95% are not uncommon. It may be noted that the Pelton wheel is best suited to conditions where the available head of water is great and the flow rate is comparatively small.

For example, with a head of 100 m and a flow rate of 1 m³/s, a Pelton wheel running at some 250 rev/min could be used to develop about 900 kW. The same water power would be available if the head were only 10 m and the flow were 10 m³/s, but a different type of turbine would then be needed.

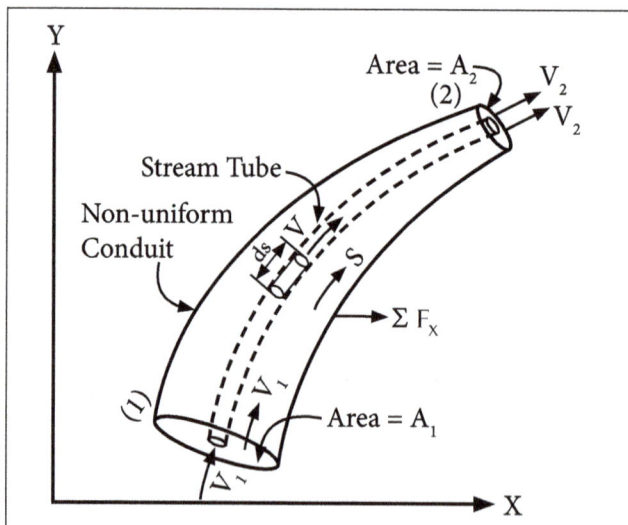

Momentum equation.

The impulse-momentum principle which is the basis of momentum equation may be stated as follows. The impulse of a force acting on a body is equal to the change of momentum produced in the body. Impulse is the product of force and time during which the force acts. Momentum is the product of mass and velocity. Since both impulse and momentum are vectors, a direction should be chosen to write the impulse-momentum principle. In the x-direction,

$$F_x \, dt = d \, (mv)x$$

Consider steady incompressible fluid flow through a non-uniform conduit shown in figure Let V_I and V_2 be the average velocities. Let v_1 and v_2 be the fluid velocities at the two end sections of a stream tube located within the conduit.

Consider a small cylindrical element of the stream tube of length ds and area of section dA. Let dF_x be the differential force acting on the element during the period of time dt.

Application of impulse-momentum principle for the element gives,

$$dF_x \, dt = d \, (\rho \, dA \, dS \, V)_x$$

The fluid density is constant and the flow is steady,

$$dF_x \, dt = \rho \, dA \, ds \, dv_x = \rho \, dA \, ds \, \frac{\partial v_x}{\partial s} \, ds$$

$$= \rho \, dA \, ds \, \frac{\partial v_x}{\partial s} \, vds = \rho \, dA \, ds \, \frac{\partial v_x}{\partial s} \, ds \, dQ \, dt$$

Where dQ is the discharge through the stream tube,

$$\therefore dF_x = \rho \frac{\partial v_x}{\partial t} \, ds \, dQ$$

Integrating this equation over the length of the stream tube,

$$F_x = \rho \, dQ \int_1^2 \frac{\partial v_x}{\partial s} \, ds$$

$$= \rho \left[(v_x)_2 - (v_x)_1 \right] dQ \qquad \qquad ...(1)$$

Where F_x is the resultant force on the stream tube in x-direction. $(v_x)_1$ and $(v_x)_2$ are the x-direction components of v_1 and v_2. The total force $\sum F_x$ on the entire conduit is obtained by integration or summation of all forces in Equation (1),

$$\therefore \sum F_x = \rho Q \left[(v_x)_2 - (v_x)_1 \right] \qquad \qquad ...(2)$$

Where $Q = V_1 A_1 - V_2 A_2$ = constant discharge through the conduit and $(v_x)_1$ and $(v_x)_2$ are the x-direction components of velocities through the two end sections. Equation (2) is called the momentum equation for x-direction.

4.1.1 Impact of Jet on Stationary and Moving Vanes (Flat and Curved)

The liquid comes out in the form of a jet from the outlet of a nozzle which is fitted to a pipe through which the liquid is flowing under pressure. The following cases of the impact of jet, i.e. the force exerted by the jet on a plate will be considered considered:

Force Exerted by the Jet on a Stationary Plate

Plate is vertical to the jet:

- Plate is inclined to the jet.
- Plate is curved.

Force exerted by the jet on a stationary vertical plate Consider a jet of water coming out from the nozzle strikes the vertical plate.

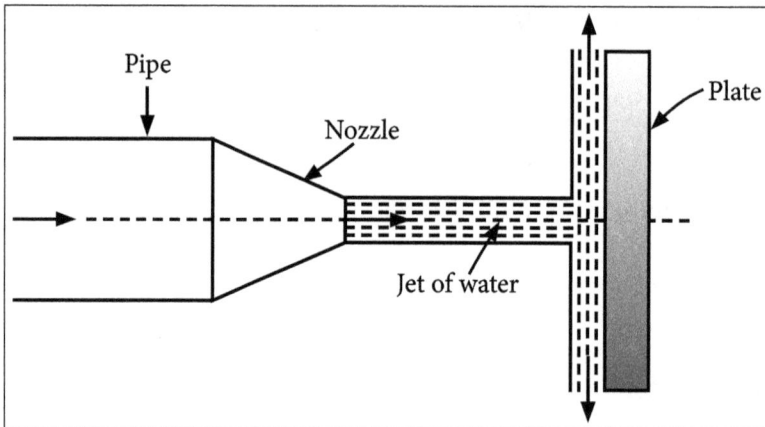

V = velocity of jet

d = diameter of the jet

a = area

x = Section of the jet

Force Exerted by the Jet on the Plate in the Direction of Jet

F_x = Rate of change of momentum in the direction of force

= initial momentum – final momentum / time

= mass x initial velocity – mass x final velocity / time

= mass/time (initial velocity – final velocity)

=mass/ sec x (velocity of jet before striking–final velocity of jet after striking)

$= \rho a V (V - o)$

$= \rho a V^2$

Force Exerted by the Jet on the Moving Plate

Force on flat moving plate in the direction of jet Consider Consider, a jet of water strikes strikes the flat moving plate moving with a uniform uniform velocity away from the jet,

V = Velocity of jet

a = Area of x-section of jet

U = Velocity of flat plate

Relative velocity of jet w.r.t plate = V – u

Mass of water striking/ sec on the plate = ρa (V - u)

Force exerted by jet on the moving plate in the direction of jet,

F_x = Mass of water striking/ sec x [Initial velocity – Final velocity]

$= \rho a (V - u) [(V - u) - o]$

$= F_x = \rho a (V - u)^2$

In this case, work is done by the jet on the plate as the plate is moving, for stationary stationary plate the w.d is zero.

Work done by the jet on the flat moving plate = Force × Distance in the direction of force/ Time,

$= W.d = \rho a (V - u)^2 \times u$

Force exerted by jet of water on single moving plate (Flat or curved) is not feasible one, it is only theoretical one.

Let,

V = Velocity of jet.

a = Area of x -section section of jet.

u = Velocity of vane In this, mass of water coming out from the nozzle/s is always in constant with plate. When all plates are considered.

Mass of water striking/s w.r.t plate = ρaV

Jet strikes strikes the plate with a velocity velocity = V – u

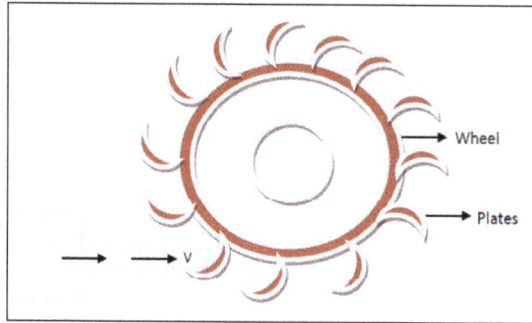

Nozzle.

Force exerted by the jet on the plate in the direction of motion of plate = Mass/sec x (Initial velocity – Final velocity),

$$F_x = \rho a V (V - u)$$

Work done by jet on the jet on the series of plate/sec,

$$\text{Work Done} = F_x \times u$$

$$\text{Work Done} = \rho a V (V - u) \times u$$

$$\text{K.E of jet/sec} = \frac{1}{2} mV^2$$

$$= \frac{1}{2} (\rho aV) V^2$$

$$\text{K.E of jet/sec} = \frac{1}{2} \rho aV^3$$

$$\text{Efficiency}, \eta = \frac{\text{Work done by jet/s}}{\text{K. E by jet/s}}$$

$$= \frac{\rho a V (V - u) \times u}{\frac{1}{2} (\rho aV) V^2}$$

$$\eta = \frac{2u (V - u)}{V^2}$$

Condition for Max Efficiency,

Efficiency is maximum when, $\dfrac{d\eta}{du} = 0$

$$\dfrac{d}{du}\left[\dfrac{2u(V-u)}{V^2}\right] = 0$$

$$\dfrac{d}{du}\left[\dfrac{2uV - 2U^2}{V^2}\right] = 0$$

$$\left[\dfrac{2V - 4u}{V^2}\right] = 0$$

$$u = \dfrac{V}{2}$$

Put the values of u in $\eta = \dfrac{2u(V-u)}{V^2}$

$$\eta_{max} = \dfrac{2\dfrac{V}{2}\left(V - \dfrac{V}{2}\right)}{V^2}$$

$$\eta_{max} = \dfrac{1}{2} = 50\%$$

4.2 Pumps and its Types

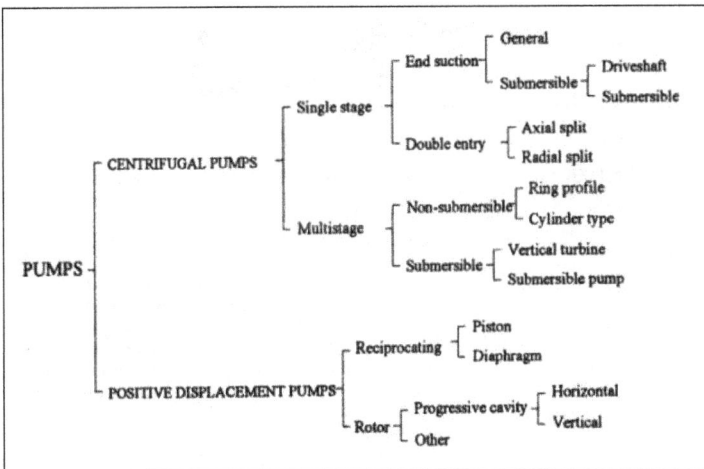

Classification of pumps for irrigation.

A simplified classification of pumps for irrigation. Most centrifugal pumps are commonly used for irrigation, but as far as the positive displacement pumps are concerned, it is only the progressive cavity rotary type pumps that are sometimes used for irrigation. The latter entails a spiral shaped rotor and horizontal and vertical types are being marketed in the RSA. These types of pumps, which can very well be commonly used for irrigation.

A Centrifugal Pump

A centrifugal pump is a roto dynamic pump that uses a rotating impeller to increase the pressure and flow rate of a fluid (Friedrichs and Kosyn 2000; Gulich 2008). Centrifugal pump are most common type of pump used to move liquids through a piping system. The fluid enters the pump impeller along or near to the rotating axis and it is accelerated by the impeller, flowing radially outward or axially into a diffuser or volute chamber, from where it exits into the downstream piping system. Centrifugal pump are typically used for large discharge through smaller heads. Centrifugal pump are often associated with the radial-flow type. However, the term "centrifugal pump" can be used to describe all impeller type roto dynamic pumps including the radial, axial and mixed-flow variations.

Positive Displacement Pump

Rotary-Type

Positive displacement rotary pump can move the fluid by using rotating mechanism that creates a vacuum that captures and draws in the liquid. Rotary positive displacement pump can be classified into two main types: Gear pumps - a simple type of rotary pump where the liquid is pushed between two gears. Rotary vane pumps - similar to scroll compressors, these pumps have a cylindrical rotor encased in a similar shaped housing. As the rotor orbits, the vanes trap fluid between the rotor and the casing, drawing the fluid through the pump.

Reciprocating-Type

Positive Displacement Pump Reciprocating pump move the fluid using one or more oscillating pistons, plungers or membranes (diaphragms), while valves restrict fluid motion to the desired direction. Pump in this category are simple with one cylinder or more. They can be either single-acting with suction during one direction of the piston 36 motion and discharge on the other or double-acting with suction and discharge in both directions.

Centrifugal pump uses the centrifugal force to pump the fluid to a certain head. The centrifugal pump is classified based on:

1. Type of Casing

According to the type of casing, centrifugal pump have two basic types:

Volute Casing

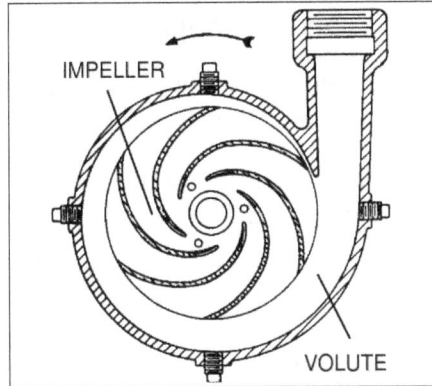
Volute casing.

In order to increase the pressure head, the velocity of fluid needs to decrease. It is done by gradually increasing the area of the casing from the impeller out let known as volute casing.

Vortex Casing

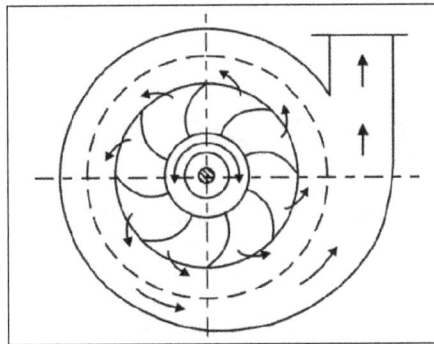
Vortex casing.

If a circular chamber known as vortex is introduce between the impeller and chamber, then the casing is known as vortex casing. Its main function is to convert kinetic energy into pressure energy.

2. Working Head

Each pump can pump up a fluid to a certain height so following are the types of centrifugal pump according to the working head:

- Low lift centrifugal pump: They pump are capable of working against heads up to 15m.

- Medium lift centrifugal pumps: They are basically used against the heads as high as 40m.

- High lift centrifugal pumps: They are used to deliver liquids at heads above 40m.

3. Liquid Handled

According to the type of liquid pumped, centrifugal pump is classified into three types:

- Pure liquid: When pure liquid is to pumped, the centrifugal pump with the closed impeller are used because they have better guidance and high efficiency.

- Little impure liquid: When liquid have a little impurity, then centrifugal pump with semi open impellers are used.

- Liquid with solid matter: When sewage, paper pulp, water containing sand or grit is to be pumped, then pump with open impeller is used.

4. Number of Impellers Per Shaft

The following are the classification based on the number of pumps:

- Single stage centrifugal pump: It has only one impeller attached to the shaft. They are used in the places where low head and low discharge rate is required.

- Multi-stage centrifugal pump: It has more then one impeller attached to a single shaft. They are used in place where high head and high discharge is required.

4.3   Centrifugal Pumps: Main Components

The following are the components of centrifugal pump. They are shown in figure.

1. Impeller

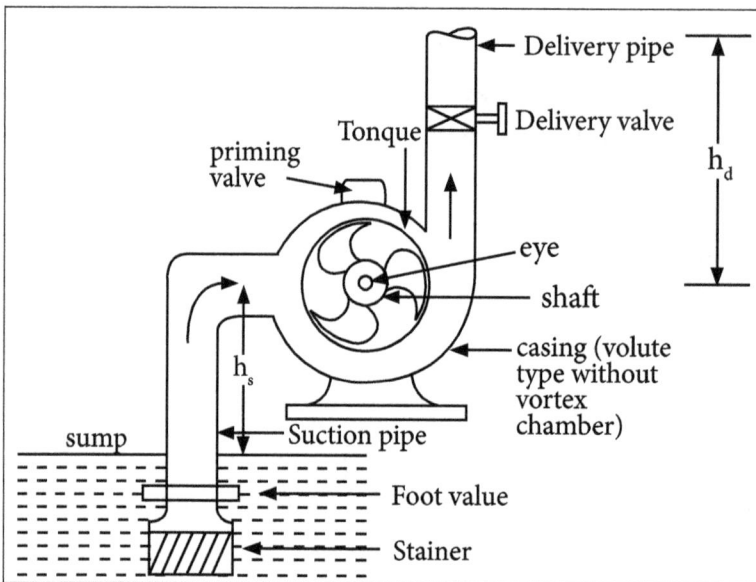

Components of Centrifugal Pump.

The rotating wheel of a centrifugal pump which is connected with motor shaft is known as impeller. It consists of a series of backward curved vanes.

In closed impeller blades or vanes are placed in between two circular disc. In open impeller no disc will be there. But blades are fitted on the periphery of shaft. In semi open impeller only one disc will be there. On eye side no disc will be there.

2. Casing

It is an air tight chamber which surrounds the impeller. It has provisions to support bearing. It is designed in such a way that the kinetic energy of the water discharged at the outlet of the impeller is converted into pressure energy before the water leaves these casing and enters the delivery pipe. The following are the types of casing.

(i) Volute champer type (ii) Vortex chamber type (iii) Diffuser pump/casing with guide blades.

i. Volute Chamber Type

This type of casing is of spiral form and has a sectional area which increases uniformly from the tongue to the delivery pipe as in figure. Velocity of whirl remains constant throughout the volute chamber of all cross sections.

ii. Vertex Chamber Pump

In a vortex chamber shown in figure a uniformly increasing area is provided between the impeller outer periphery and the volute casing. This increase in area leads to increase in pressure and decrease in velocity in the outward direction.

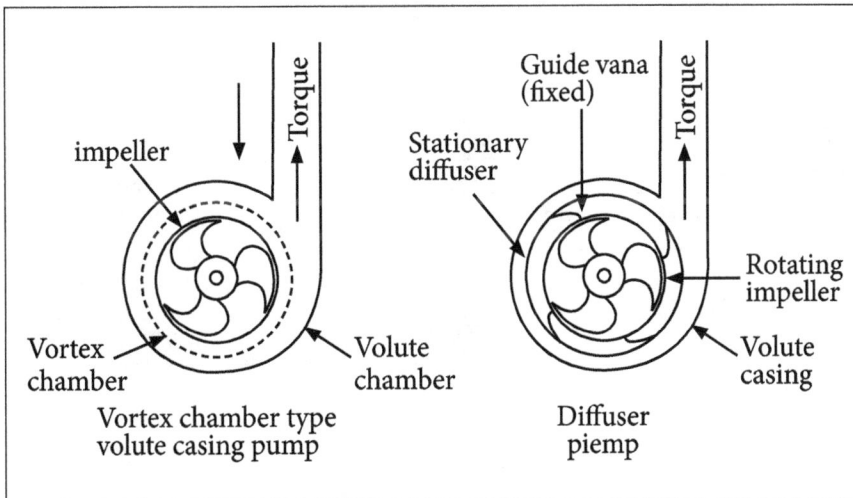

Types of Casing.

iii. Diffuser Pump/Casing Guide Blades

The guide vanes are arranged at the outlet of the impeller vanes as shown in figure. Water enters the guide vanes without shock. As the guide vanes are of enlarging cross section area the velocity of water decreases and pressure increases. Since these vanes provide better guidance to flow eddy losses are reduced which increases the efficiency.

3. Suction Pipe with a Foot Value and a Strainer

A pipe whose one end is connected to the inlet of the pump and other end submerged into water in a sump is known as suction pipe. A foot value which is a non-return valve is fitted at the lower end of the suction pipe. The foot value opens only in the upward direction. A strainer is also fitted at the lower end the suction pipe to remove strains.

4. Delivery Pipe

A pipe whose one end is connected to the outlet of the pump and the other end delivers the water at a required height is known as delivery pipe.

4.3.1 Working Principle

Priming is the first operation in which the suction pipe, casing of the pump and the portion of the delivery pipe up to delivery valve are completely filled with liquid which is to be pumped so that all the air or gas or vapor from this portion of the pump is driven out and no air pocket is left.

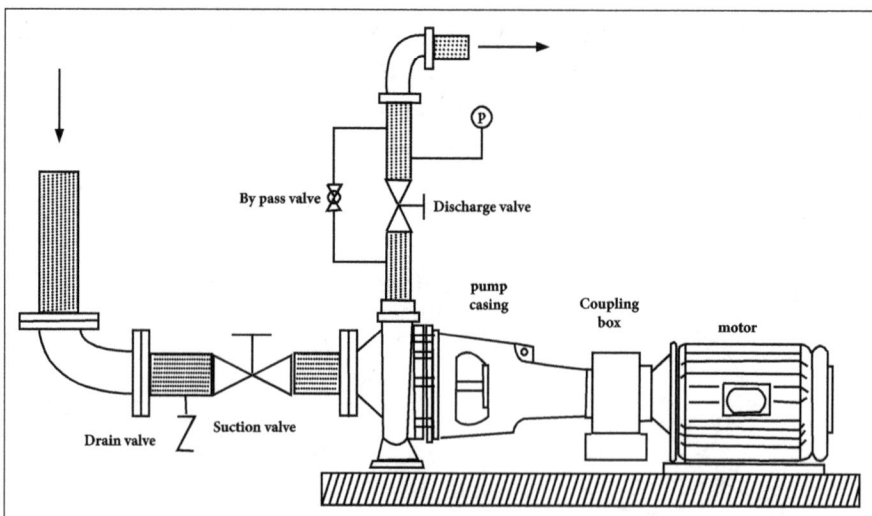

Centrifugal pump.

The delivery valve being in closed position, the electric motor is started to rotate the impeller. When the delivery valve is opened, the liquid is made to flow in an outward radial direction thereby having the vanes of the impeller, at the outer circumference with high velocity and pressure and thereby at the eye of the impeller, vacuum is formed which causes the water to enter the casing or impeller from the sump. Suitable casing is provided in order to convert the velocity energy into pressure energy with gradual reduction of velocity so that the loss of energy due to eddy formation is minimum.

4.3.2 Multi Stage Pumps

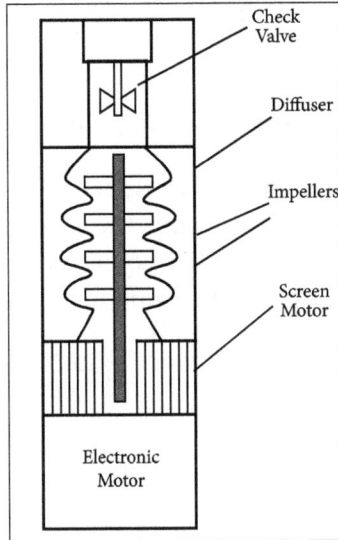

Multistage pump.

- To increase high head a number of impellers are mounted in series or on the same shaft.

- To increase the discharge under constant head which is achieved by parallel arrangement of impellers.

- The loss due to friction is less leakage reduced stress.

- Axial thrust is eliminated and higher suction lift is possible.

- The noise is very less compare then reciprocating pump.

- Due to high pressure head, the liquid can be lifted to a high level.

- This pump pushes water to surface instead of striking water from ground.

- It is placed inside water and acts like a multistage centrifugal pump operating in vertical position.

- Water enters the pump through the screen and is subjected to great centrifugal force by the speed of impeller.

- The water losses, their kinetic energy in the diffuser where conversion of kinetic to pressure energy takes place.

- A strainer with foot valve is also connected at the end of a suction pipe to prevent foreign particles in order to avoid from entering the suction pipe and also to reduce suction losses.

- The water is pumped by entirely the centrifugal action of the pump in terms of impeller rotation.

- The pump is driven by either electric motor or internal combustion engine which is coupled to pump.

4.3.3 Performance and Characteristic Curves

In the volute of the centrifugal pump, the cross section of the liquid path is greater than the impeller, and in an ideal frictionless pump, the drop from the velocity V to the lower velocity is converted according to Bernoulli's equation, to an increased pressure. This is the source of the discharge pressure of the centrifugal pump.

If the speed of the impeller is increased from N_1 to N_2 rpm, its flow rate will increase from Q_1 to Q_2 as,

$$\frac{Q_1}{Q_2} = \frac{N_1}{N_2}$$

The head developed (H) will be proportional to the square of the quantity discharged, so that,

$$\frac{H_1}{H_2} = \frac{Q_1^2}{Q_2^2} = \frac{N_1^2}{N_2^2}$$

The power consumed (W) will be the product of H and Q and, therefore,

$$\frac{W_1}{W_2} = \frac{Q_1^3}{Q_2^3} = \frac{N_1^3}{N_2^3}$$

These relationships form only the roughest guide to the performance of centrifugal pumps.

Operating Characteristics of Pumps

Pumping mechanisms can be broken into two major categories positive displacement pumps and kinetic pumps. The positive displacement of fluids is the principle behind positive displacement pumps. The most common types of positive displacement pumps are reciprocating and rotary pumps. a. Reciprocating pumps are similar in theory to reciprocating engines.

A piston is used to move fluid into a cylinder on the down stroke then move the fluid out of the cylinder on the upstroke. Reciprocating pumps provide high suction lift and high pressure for small quantities of flow as the pistons move up and down. Rotary pumps use a screw-type device to trap fluids in the cylinder and force it out of the chamber. This type of pump is used with gasoline or other low viscosity fuels when high suction lift is required. The rotary pump provides a smooth, even flow.

Centrifugal pumps depend on centrifugal force for their operation. Centrifugal force acts on the body moving in a circular path tending to force it farther away from the axis or center point of the circle described by the path of the rotating body.

In a centrifugal pump the power plant turns an impeller which creates centrifugal force in the pump housing and forces fuel out of the pump. Major advantages to centrifugal pumps are fewer moving parts smooth, non-pulsating flow and a much higher capacity than positive displacement pumps. One of the disadvantages of centrifugal pumps is that it has are relatively lower head capacity than positive displacement pumps and the head capacity is based on the design of the impeller.

Pumps are chosen for particular requirement. The requirements are not constant as per example the pressure required for flow through a piping system. As flow increases the pressure required increases. In the case of the pump as flow increases, the head decreases. The operating condition will be the meeting point of the two curves representing the variation of head required by the system and the variation of head of the pump. This is shown in Figure.

The operating condition decides about the capacity of the pump or selection of the pump. If in a certain setup there is a need for increased load; either a completely new pump may be chosen. This may be costlier as well as complete revamping of the setup. An additional pump can be the alternate choice. If the head requirement increases the old pump and the new pump can operate in series.

In case more flow is required the old pump and the new pump will operate in parallel. There are also additional advantages in two pump operation. When the Pump-load characteristics load is low one of the pump can operate with a higher efficiency when the load increases then the second pump can be switched on thus improving part load efficiency.

5

Hydraulic Turbines

5.1   Hydraulic Turbines: Classification of Turbines

A hydraulic turbine is a prime mover that uses the energy of flowing water and converts it into the mechanical energy. This mechanical energy is used in running an electric generator which is directly coupled to the shaft of the hydraulic turbine. From this electric generator, we get electric power that can be transmitted over long distances by means of transmission lines and transmission towers. The hydraulic turbines are also called as 'water turbines' as the fluid medium used in them is water.

The main classification depends upon the type of action of water on the turbine. They are:

- Reaction Turbine.

- Impulse turbine.

1. In reaction turbines the available potential energy is progressively converted in the turbine rotors and the reaction of the accelerating water causes the turnings of the wheel. These are further divided as radial flow, mixed flow and axial flow machines. Radial flow machines are found suitable for the moderate levels of potential energy and medium quantities of flow. The axial machines are suitable for low levels of the potential energy and large flow rates. The potential energy available is generally denoted as the "head available". With this terminology, plants are designated as "high head", "medium head" and "low head" plants.

2. In the case of impulse turbine all the potential energy is converted to kinetic energy in nozzles. The impulse provided by the jets is used to turn the turbine wheel. The pressure inside the turbine is atmospheric. This type is found suitable when available potent ideal energy is high and flow available is comparatively low. Classification of turbines based on specific speed:

- High specific speed (300 to 1000). Example: Kalpan Turbine.

- Medium specific speed (60 to 400). Example: Francis Turbine.

- Low specific speed (10 to 35). Example: Pelton Wheel.

The various types of turbines are:

1. According to the action of water on moving blades:

- Reaction low head.

- Impulse high head.

2. According to the direction of flow of water:

- Radial flow.

- Mixed flow eg.: Francis.

- Tangential flow eg.: Pelton Wheel.

- Axial flow eg.: Kaplan.

3. According to the specific speed:

- High specific speed NS > 400.

- Medium specific speed NS = 80 to 400.

- Low specific speed NS < 70.

Reaction Turbine: When at the inlet of the turbine the water possesses kinetic energy as well as pressure energy the turbine is known as reaction turbine.

Impulse Turbine: When the inlet of the turbine the energy available is only kinetic energy the turbine is known as impulse turbine. The water flows over the vanes the pressure is atmospheric from inlet to outlet of the turbine.

Inward Radial Flow Turbine: When the water flows from outwards to inwards radially.

Outward Radial Flow Turbine: When the water flows radially from inwards to outwards.

Axial Flow Turbine: When the water flows through the runner along the direction parallel Tangential.

Flow Turbine: When the water flows along the tangent of the runner.

Mixed Flow Turbine: When the water flows through the runner in the direction but leaves in the direction parallel to axis of rotation of the runner.

Radial Flow Turbine: When the water flows in the radial direction through the runner to the axis of rotation of runner.

5.1.1 Working Principle

If at the inlet of turbine, the energy available is only kinetic energy, the turbine is known as the impulse turbine. As the water flows over the vanes, the pressure is atmospheric from inlet to the outlet of the turbine. In the impulse turbine, all the potential energy of water is converted into kinetic energy in the nozzle before striking the turbine wheel buckets. Thus an impulse turbine requires high head and low discharge at the inlet. The water as it flows over the turbine blades may be at the atmospheric pressure. The impulse turbine can be of radial flow or tangential flow type.

If at the inlet of the turbine, the water possesses kinetic energy as well as the pressure energy, the turbine is called as the reaction turbine. As the waters flows through the runner, it is under pressure and the pressure energy goes on changing into kinetic energy. The runner is completely enclosed in air tight casing and the runner along with the casing is completely full of water.

Suppose if the water flows along the tangent of the runner, the turbine is called as tangential flow turbine. If the water flows in the radial direction through runner, the turbine is known as the radial flow turbine. If the water flows from outwards to inwards, radially the turbine is known as the inward radial flow turbine, While, if the water flows radially from inwards to outwards, the turbine is called as the outward radial flow turbine. If the water flows through the runner along the direction parallel to axis of rotation of the runner, the turbine is known as the axial flow turbine. If the water flows through the runner in radial direction but leaves in the direction parallel to axis of rotation of the runner, the turbine is known as the mixed flow turbine.

5.1.2 Efficiency Calculation and Design Principles for Pelton Wheel

A Pelton wheel turbine has a rotor at the periphery which is mounted on an equally spaced double hemispherical (or) double ellipsoidal buckets. The water is transferred from a high head source through pen stock that is fitted with a nozzle through which the water flows out as a high jet.

A needle spear moving inside the nozzle controls water flows through the nozzle at the same time. It gives a smooth flow with negligible energy loss. All the potential energy available is converted into the kinetic energy before the jet strikes the buckets of the runner. The pressure all over the wheel is constant and equal to the atmosphere.

The Pelton wheel is provided with a casing. The function of casing is to prevent the splashing of water and to discharge water to the tail race.

The nozzle is completely closed by noting the spear in the forward direction of a amount of water striking the runner is reduced to zero but the runner due to inertia continues revolving for a long time. In order to bring the runner to rest in a short time, a nozzle [brake] is provided which directs the jet of water on the buckets. This jet of water is called breaking jet.

Speed of the turbine is kept constant by a governing mechanism, that automatically regulates the quantity of water flowing through the runner in accordance with any variation of load.

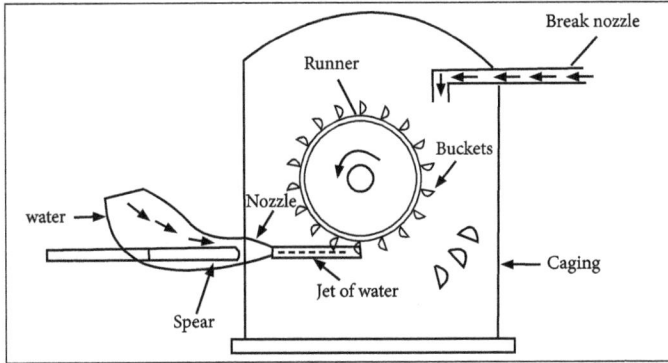

Pelton wheel.

Maximum power,

$$P = \rho Q U (V_1 - U)\left[1 + k\cos(180 - \theta)\right]$$

Hydraulic Efficiency,

$$\eta_h = \frac{2U(V_1 - U)\left[1 + k\cos(180 - \theta)\right]}{V_1^2}$$

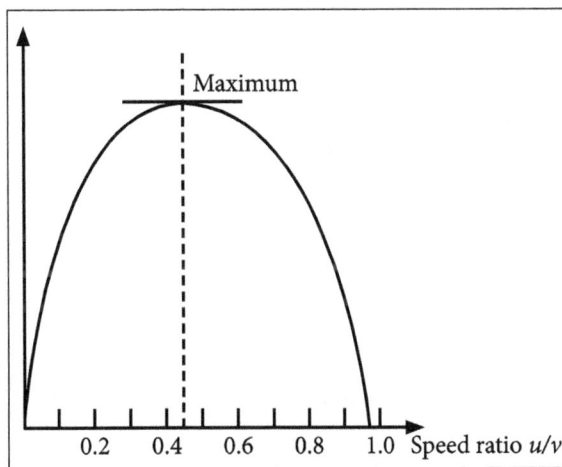

Velocity at maximum hydraulic efficiency is given by,

$$U = \frac{V_1}{2}$$

Maximum hydraulic efficiency is given by,

$$\eta_h = \frac{1 + k \cos(180 - \theta)}{2}$$

Power at maximum hydraulic efficiency is given by,

$$P_{max} = \rho Q \left(\frac{V_1^2}{4}\right)\left[1 + k \cos(180 - \theta)\right]$$

Overall efficiency,

$$\eta_o = \frac{P}{\rho g Q H}$$

Hydraulic efficiency,

$$\eta_h = \frac{V_{w1} U_1 - V_{w2} U_2}{g H}$$

Volumetric efficiency,

$$\eta_v = \frac{Q_a}{Q}$$

Mechanical Efficiency

It is defined as the ratio of the power obtained from the shaft of the turbine to the power developed by the runner. These two power differ by the amount of mechanical losses. bearing friction.

$$\eta_m = \frac{\text{Power available of the turbine shaft}}{\text{Power developed by turbine runner}}$$

$$= \frac{P}{\omega Q_a \cdot H_r}$$

Value of mechanical efficiency for a Pelton wheel usually lies between 97 to 99%. The mechanical efficiency will be more if the mechanical loss is less and capacity of the unit is high.

Working Proportions

The diameter of the wheel, its angular speed, and the jet diameter and bucket dimensions are all quantities that are fairly rigidly related in Pelton wheels. The relations among these quantities have been determined partly on theoretical grounds and partly empirically. Some of the most important relations are as follows:

- Jet velocity $V = (0.96 \text{ to } 0.98)(2gH)^{1/2}$.

- Wheel rotational speed is $(0.44 \text{ to } 0.46)(2gH)^{1/2}$ at pitch diameter.

- Ratio of wheel pitch diameter to jet diameter = D/d, varies between 14 and 16.

In extreme cases, it can be as low as 6:

- Axial width of the buckets is 2.8 to 3.2 times the jet diameter. Sometimes, it may even range upto 4.0d.

- The number of buckets to be mounted on the wheel is determined by the requirements that there are just sufficient buckets to utilize the kinetic energy of the water completely, without splashing and a consequent wastage of water and its energy. Based on this requirement, Jagdish Lal provides a theory that predicts too few buckets. An approximate empirical equation provided by Tygun is,

$$Z = D/(2d) + 15$$

Since m = D/d is about 15, this equation predicts Z = 20-22, thus making it almost constant for all Pelton wheels. Nechleba. On the other hand, provides an approximate method of determining Z and gives the following table for the variation of the number of buckets with the jet-diameter ratio.

Table: Appropriate number of buckets for a Pelton wheel:

m	6	8	10	15	20	25
Z	17 - 21	18 - 22	19 - 24	22 - 27	24 - 30	26 - 33

It is possible to derive an expression relating the jet diameter ratio and the specific speed of the Pelton wheel in terms of the jet velocity coefficient, speed ratio and other parameters. Let the jet velocity coefficient be Ci so that $V_1 = C_1 (2gH)^{1/2}$. Then if Q is the volume flow rate of water emerging from the nozzle, one can write for the jet diameter the equation,

$$n(\pi d^2/4)V_1 = Q \text{ or } d^2 = 4Q/\left[\pi Cjn(2gH)^{1/2}\right]$$

Where n is the number of jets used for the flow. Since the total power output is P = (turbine efficiency) x (theoretical power), one gets,

$$P = (\eta \rho g / g_c)(QH/75) = (\rho n \eta g/g_c)(\pi d^2/300)\left[Cj(2gH)^{1/2} H\right]$$

Also,

$$N = 60u(\pi D) = 60\phi(2gH)^{1/2}/(\pi D)$$

$$N_s = \left[60 \; \phi d^1 \, D\right](2g)^{3/4}\left[\eta\rho nC_1 g/(300\pi g_c)\right]^{1/2}$$

$$= 570 \; \phi(\eta nC_1)^{1/2}/m$$

Assuming $\phi = 0.45$, $C_j = 0.97$ and $\eta = 0.89$, one can calculate the expression,

$$N_s = 240\sqrt{n}/m$$

Thus, the specific speed of a Pelton wheel is inversely proportional to the diameter ratio, m. Since m is approximately 15, the efficiency of a single-jet Pelton wheel is maximum when it has a specific speed of about 16. If a turbine of higher specific speed is desired, multiple jets have to be used to maintain the jet-diameter ratio. So far, Pelton wheels using a maximum of six jets have been built.

Problems

1. The mean velocity of the buckets of the Pelton wheel is 10 m/s. The jet supplies water at 0.7 m³/s at a head of 30 m. The jet is deflected through an angle of 160° by the bucket. Let us determine the hydraulic efficiency. $C_v = 0.98$.

Solution:

Given,

Mean velocity = 10 m/s

$$Q = 0.7 \text{ m}^3/\text{s}$$

Deflected angle = 1600

Formula to be used,

$$V_1 = C_v\sqrt{2gH}$$
$$V_{r1} = V_1 - u_1$$
$$V_{w2} = V_{r2}\cos\phi - u_2$$
$$\eta_h = \frac{2[V_{w1} + V_{w2}]}{V_1^2} \times u$$
$$Q = 0.7\,\text{m}^3/\text{s}$$

Angle of deflection = 160°

$$\therefore \text{Angle}, \phi = 180° - 160° = 20°$$

Coefficient of velocity, $C_v = 0.98$

$$H = 30\,m$$

Velocity of jet, $V_1 = C_v \sqrt{2gH}$

$$= 0.98 \sqrt{2 \times 9.81 \times 30}$$

$V_1 = 23.77\,m/s$

$V_{\omega 1} = V_1 = 23.77\,m/s$

$V_{r1} = V_1 - u_1 = 23.77 - 10$

$V_{r1} = 13.77\,m/s$

$V_{r2} = V_{r1} = 13.77\,m/s$

$V_{\omega 2} = V_{r2}\cos\phi - u_2$

$$= 13.77\cos 20° - 10$$

$V_{\omega 2} = 2.94\,m/s$

The hydraulic efficiency of the turbine,

$$\eta_h = \frac{2\left[V_{\omega 1} + V_{\omega 2}\right]}{V_1^2} \times u$$

$$= \frac{2\left[23.77 + 2.94\right]}{23.77 \times 23.77} \times 10$$

$\eta_h = 0.9454\,(or)\;94.54\%$

$u = u_1 = u_2 = 10\,m/s$

2. A Pelton wheel supplies water from reservoir under a gross head of 112 m and the friction losses in the pen stock amounts to 20 m of head. The water from pen stock is discharged through a single nozzle of diameter of 100 m at the rate of 0.30 m³/s. Mechanical losses due to friction amounts to 4.3 KW of power and shaft power available is 208 KW. Let us determine:

- Velocity of Jet.

- Water Power at Inlet to Runner.

- Power Loss in Nozzles.

- Power Lost In Runner Due To Hydraulic Resistance.

Solution:

Given,

$$H_g = 112 \text{ m}$$

$$h_f = 20 \text{ m}$$

$$d = 100 \text{ mm} = 0.1 \text{ m}$$

$$Q = 0.30 \text{ m}^3/S$$

Formula to be used,

$$V_1 = \frac{Q}{a} = \frac{0.30}{0.00785}$$

$$V_1 = 38.197 \text{ m/s}$$

$$W.P = \frac{\rho g Q H}{1000}$$

$$= \frac{1000 \times 9.81 \times 0.30 \times 92}{1000}$$

$$W.P = 270.756 \text{ kW}$$

Power of jet + Power lost in nozzle.

Power at the shaft + Power lost in nozzle + Power lost in runner + Power lost due to mechanical resistance.

Power lost in Mechanical friction = 0.45 kW

Shaft power = 208 kW

1. Velocity of Jet

Total head, $H = H_g - h_f$

$$= 112 - 20$$

$$H = 92 \text{m}$$

Area of Jet, $a = \frac{\pi}{4} d^2$

$$= \frac{\pi}{4} \times (0.1)^2$$

$$a = 0.00785 \text{ m}^2$$

$$Q = a \times V_1$$

$$V_1 \frac{Q}{a} = \frac{0.30}{0.00785}$$

$$V_1 = 38.197 \text{ m/s}$$

2. Water Power at Inlet to Runner

$$W.P = \frac{\rho g Q H}{1000}$$

$$= \frac{1000 \times 9.81 \times 0.30 \times 92}{1000}$$

$$W.P. = 270.756 \text{ kW}$$

Power corresponding to kinetic energy of Jet in Kw,

$$= \frac{1}{2} \frac{mV^2}{1000}$$

$$= \frac{1}{2} \frac{(\rho \times a V_1) V_1^2}{1000}$$

$$= \frac{1}{2} \frac{\rho \, Q V_1^2}{1000}$$

$$= \frac{1}{2} \times \frac{1000 \times 0.30 \times (38.197)^2}{1000}$$

$$= 218.85 \text{ kW}$$

3. Power Lost in Nozzle

Power at the base of nozzle = Power of jet + Power lost in nozzle.

$$270.756 = 218.85 + \text{Power lost in Nozzle.}$$

Power lost in Nozzle = 51.90 kW

4. Power Lost in Runner Due to Hydraulic Resistance

Power at the base of Nozzle = Power at the shaft +Power lost in nozzle + Power lost in runner + Power lost due to mechanical resistance,

$$270.756 = 208 + 51.9 + \text{Power lost in runner} + 4.3$$

Power lost in runner = 6.556 kW

3. A Pelton turbine having 1.6 m bucket diameter develops a power of 3600 KW at 400 rpm, under a net head of 275 m. If the overall efficiency is 88% and the coefficient of velocity is 0.97, let us determine:

- Speed Ratio.

- Discharge.

- Diameter of the Nozzle and Specific Speed.

Solution:

Given data,

$$D = 1.6 \text{ m}$$

$$SP = 3600 \text{ kW}$$

$$N = 400 \text{ rpm}$$

$$H = 275 \text{ m}$$

$$\eta_o = 88\%$$

$$C_v = 0.97$$

Formula to be used,

$$u = \frac{\pi DN}{60}$$

$$\eta_o = \frac{SP}{WP} = \frac{SP}{\dfrac{\rho g \times Q \times H}{1000}}$$

$$Q = A \times V$$

$$V_1 = C_v \sqrt{2gH}$$

$$N_s = \frac{N\sqrt{Q}}{\left(H_m\right)^{3/4}}$$

1. Speed Ratio

$$u = \frac{\pi DN}{60}$$

$$u = \frac{\pi \times 1.6 \times 400}{60}$$

$$u = 33.51 \text{ m / s}$$

$$u = \phi \sqrt{2gH}$$

$$33.51 = \phi \sqrt{2 \times 9.81 \times 275}$$

$$\phi = 0.45$$

2. Discharge

$$\eta_o = \frac{SP}{WP} = \frac{SP}{\dfrac{\rho g \times Q \times H}{1000}}$$

$$0.88 = \frac{3600}{\dfrac{100 \times 9.81 \times Q \times 275}{1000}}$$

$$Q = 1.516 \text{ m}^3 / \text{s}$$

3. Diameter of Nozzle

$$Q = A \times V$$

$$V_1 = C_v \sqrt{2gH} = 0.97 \times \sqrt{2 \times 9.81 \times 275}$$

$$V_1 = 71.25 \text{ m / s}$$

$$1.516 = \frac{\pi}{4} d^2 \times 71.25$$

$$d = 0.16 \text{ m}$$

4. Specific Speed

$$N_S = \frac{N\sqrt{Q}}{\left(H_m\right)^{3/4}}$$

$$= \frac{400 \times \sqrt{1.156}}{\left(275\right)^{3/4}}$$

$$N_S = \frac{492.50}{67.53}$$

$$N_S = 7.29 \text{ rpm}$$

5.2 Francis Turbine

In Francis Turbine, water flow is radial into the turbine and exits at the turbine axially. Water pressure decreases as it passes through the turbine, imparting reaction on the turbine blades and makes the turbine rotate. Francis Turbine is the first hydraulic turbine with the radial inflow. It was designed by an American scientist James Francis.

Francis Turbine is a reaction turbine. Reaction turbines do have some primary features that differentiate from the Impulse Turbines. The major part of pressure drop occurs in the turbine itself, rather the impulse turbine where complete pressure drop takes place until the entry point and the turbine passage are completely filled by the water flow during operation.

Design of the Francis Turbine

Francis Turbine consists of a circular plate fixed to the rotating shaft that is perpendicular to its surface and passes through its center. The circular plates also have curved channels on it. The plate with channels is collectively called as the runner. The runner is encircled by a ring of stationary channels called as guide vanes. Guide vanes are housed in spiral casing known as volute.

The exit of the Francis turbine is at the center of runner plate. There is a draft tube attached to the central exit of the runner. The design parameters like curvature of channel, radius of the runner, angle of vanes and the size of the turbine as a whole depends upon the available head and the type of application altogether.

Working of Francis Turbine

Francis Turbines are basically installed with the help of their axis vertical. Water with high head enters the turbine through the spiral casing surrounding guide vanes. The water loses a part of its pressure in the volute to maintain the speed. Then water passes through guide vanes where they are directed to strike the blades on the runner at the optimum angles.

As water flows its pressure and angular momentum reduces. This reduction imparts reaction on runner and the power is transferred to turbine shaft. Turbine is operating at design conditions and the water leaves the runner in the axial direction. Water exits the turbine through the draft tube that acts as a diffuser and reduces exit velocity of the flow to recover the maximum energy from flowing water.

Power Generation using Francis Turbine

For power generation using Francis Turbine, the turbine is supplied with high pressure water that enters the turbine with the radial inflow and leaves the turbine axially through the draft tube. The energy from the water flow is transferred to shaft of turbine in the form of torque and rotation.

The turbine shaft is coupled with the dynamos or the alternators for the power generation. For quality power generation, speed of turbine must be maintained constant despite the change of loads. To maintain the runner speed constant even in the reduced load condition, the water flow rate is reduced by changing guide vanes angle.

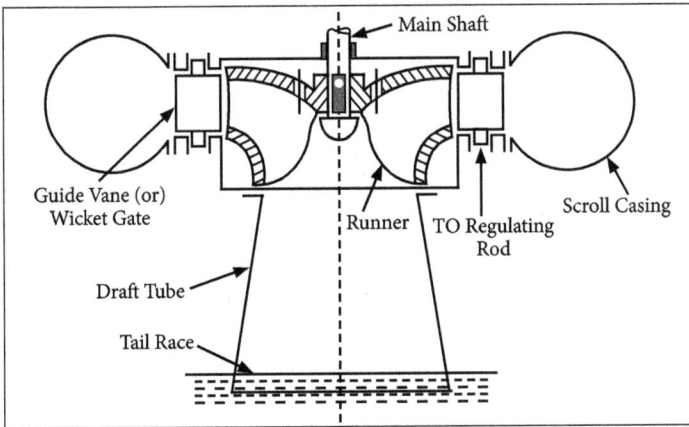

Francis turbine.

Blade Efficiency

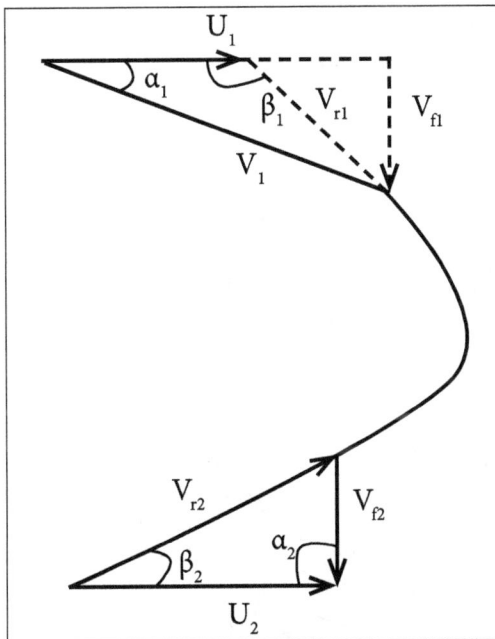

Inlet velocity triangle.

Generally the flow velocity remains constant throughout, i.e., $V_{f_1} = V_{f_2}$ and it is equal to the inlet to the draft tube.

Using Euler turbine equation,

$$E/m = e = V_{w1} U_1$$

Where, e is the energy transferred to the rotor per unit mass of the fluid.

From the inlet velocity triangle,

$$V_{w1} = V_{f1} \cot\alpha_1$$

$$U_1 = V_{f1}(\cot\alpha_1 + \cot\beta_1)$$

Hence,

$$e = V_{f1}^2 \cot\alpha_1 (\cot\alpha_1 + \cot\beta_1)$$

The loss of kinetic energy per unit mass thus becomes $V_{f2}^2/2$.

Hence, neglecting friction, the blade efficiency becomes,

$$\eta_b = e/\left(e + V_{f2}^2/2\right)$$

i.e.,

$$\eta_b = \frac{2V_{f1}^2\left(\cot\alpha_1(\cot\alpha_1 + \cot\beta_1)\right)}{V_{f2}^2 + 2V_{f1}^2\left(\cot\alpha_1(\cot\alpha_1 + \cot\beta_1)\right)}$$

Velocity Diagram

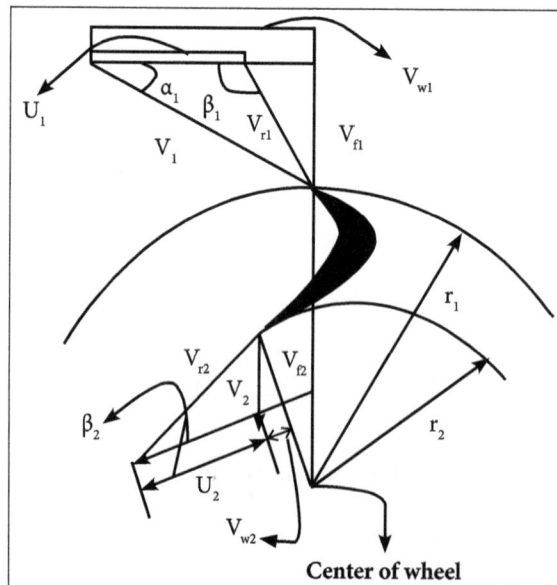

Center of wheel

The degree of reaction is defined as the ratio of pressure energy change in blades to the total energy change of fluid. This implies that it is a ratio indicating the fraction of total change in fluid pressure energy that occurs in the blades of the turbine.

Rest of the changes occurs in the stator blades of the turbines and the volute casing, as it possess a varying cross-sectional area. For instance, if the degree of reaction is given to be 50%, that means half of the total energy change of the fluid is taking place in the rotor blades and other half is occurring in stator blades. If the degree of reaction is zero, it implies that the energy changes due to rotor blades is zero leading to a different turbine design,

$$R = e^{-1/2} \left(V_{f_1}^2 - V_{f_2}^2 \right) / e$$

Now, by putting the value of 'e' from above and using $\left(V_{f_1}^2 - V_{f_2}^2 = V_{f_1}^2 \cot \alpha_2 \text{ as } V_{f_2} = V_{f_1} \right)$

$$R = 1 - \left(\cot \alpha_1 / 2 \left(\cot \alpha_1 + \cot \beta_1 \right) \right)$$

5.2.1 Kaplan Turbine

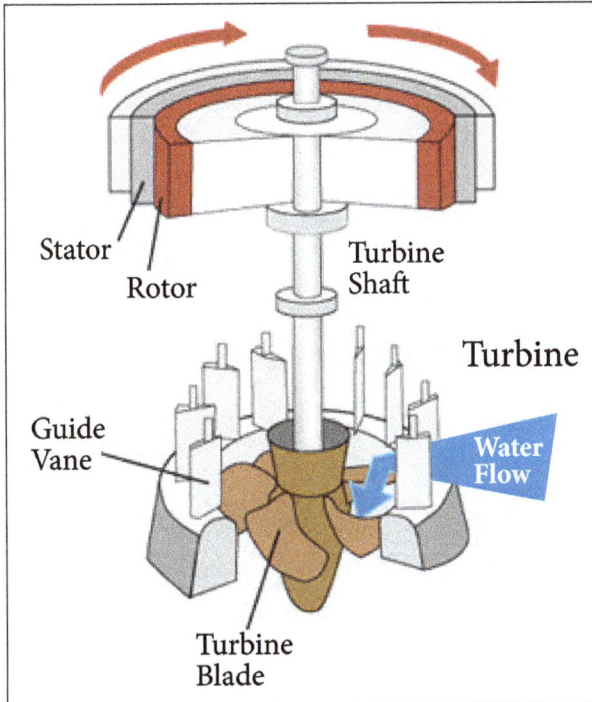

Design of Kaplan turbine.

The Kaplan turbine is one of the most widely used propeller-type water turbine that has adjustable blades. It was developed in 1913 by an Austrian professor Viktor Kaplan, by combining the adjusted propeller blades with automatically adjusted wicket gates to achieve the efficiency over a wide range of flow and the water level.

The Kaplan turbine was an evolution of the Francis turbine. This invention allowed the efficient power production in the low-head applications that was impossible with Francis turbines. Kaplan turbines are now widely used throughout the world in high-flow, low-head power production. The Kaplan turbine is an inward flow reaction turbine, that implies that working fluid varies the pressure as it moves through turbine and gives up its energy. Its design combines both the axial and radial features.

The inlet is a scroll-shaped tube which wraps around the wicket gate of the turbine. Water is directed tangentially through the wicket gate and spirals on to a propeller shaped runner causing it to the outlet .It is specially shaped draft tube that helps for decelerate the water and recovers kinetic energy.

The turbine is not needed to be at the lowest point of water flow as long as the draft tube remains with full of water. A higher turbine location will increases the suction that is imparted on the turbine blades with the help of draft tube. The resulting pressure drop will lead to cavitation.

Variable geometry of wicket gate and turbine blades allows efficient operation for a range of flow conditions. Their efficiencies are typically over 90%, but sometimes may be lower in a very low head applications.

The current areas of research includes CFD driven efficiency improvements and new designs that raises the survival rates of fish that pass through them.

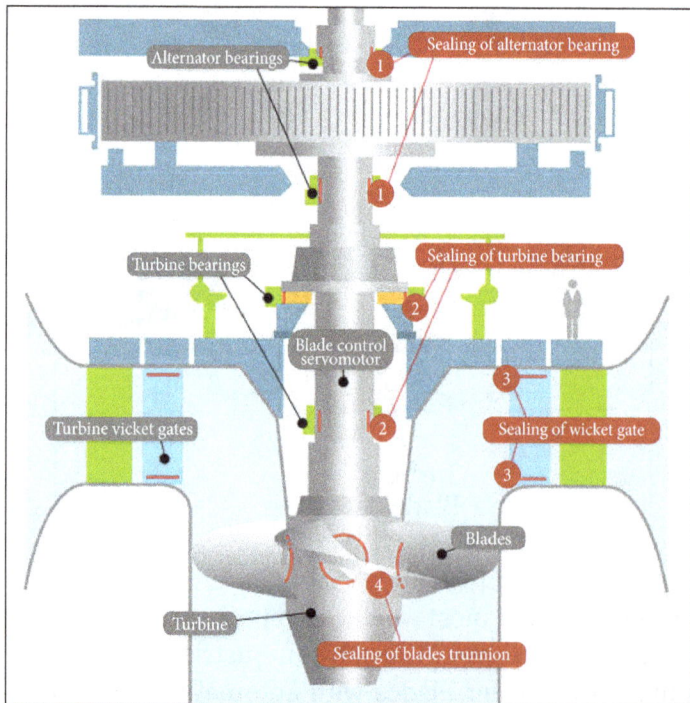

Kaplan turbine.

Kaplan turbines are widely used throughout the world for the electrical power production. They cover the lowest head hydro sites and they are especially suited for high flow conditions. Some of the inexpensive micro turbines are being manufactured for the sake of individual power production with so little two feet of head.

Kaplan turbine is a low head turbine. Large Kaplan turbines are designed individually for each site to operate at the highest possible efficiency over 90%. They are very expensive to design, manufacture and install but at the same time, they operate for decades.

The Kaplan turbine is one of the greatest developments of early 20th century. It was invented by Professor Viktor Kaplan of Austria. The Kaplan belongs to the propeller type, much like an airplane propeller. The basic difference between the Propeller and Kaplan turbines is that the Propeller turbine has a fixed runner blades but the Kaplan turbine has adjustable runner blades.

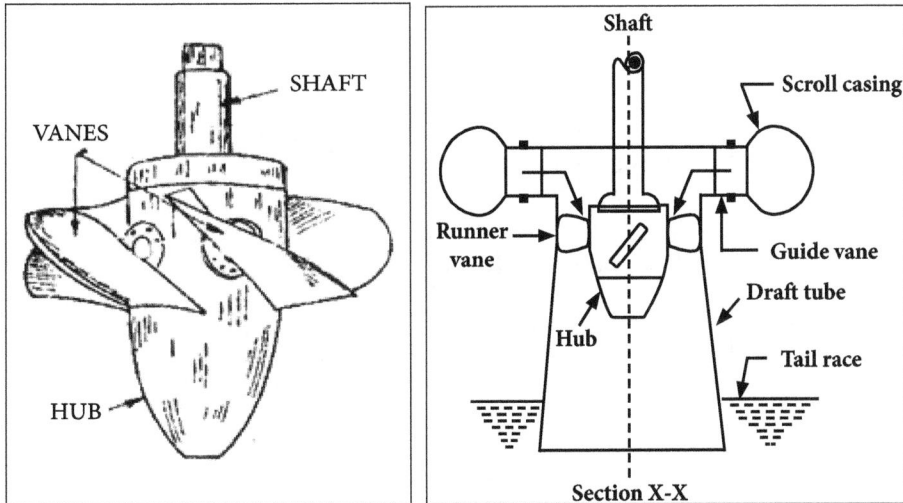

Construction of Kaplan Turbine.

It is a pure axial flow turbine. The kaplan blades are adjustable for the pitch and will handle a great variation of flow much efficiently. They are 90% or better in efficiency and they are used to place some of the old Francis type turbines.

They are really expensive. In Kaplan turbine, the runner's blades are movable. The application of Kaplan turbines are from a head of 2m to 40m.

Constructional details of the Kaplan Turbine

The main parts of Kaplan Turbine are:

- Guide vanes mechanism.

- Scroll casing.

- Draft tube.

- Hub with vanes or runner of the turbine.

1. Scroll Casing

The water from the penstocks enters into the scroll casing and then moves to the guide vanes. From the guide vanes, water turns through 90° and flows axially through runner.

2. Guide Vane Mechanism

The Guide Vanes are fixed on the Hub.

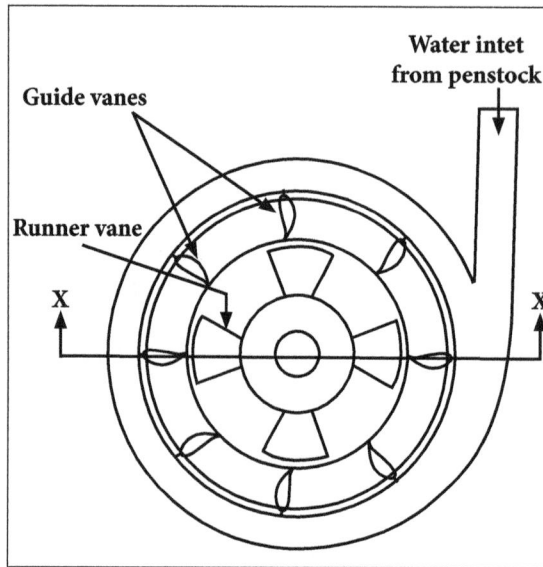

Guide Vane Mechanism.

3. Hub

For Kaplan Turbine, the shaft of the turbine is vertical. The lower end of shaft is made larger and is known as the 'Hub' or 'Boss'. The vanes are fixed on the hub and thus Hub acts as a runner for axial flow turbine.

4. Draft Tube

The pressure at the exit of runner of Reaction Turbine is usually less than the atmospheric pressure. The water at exit may not be directly discharged to the tail race. The tube or pipe of gradually increases the area and that is used for discharging water from the exit of turbine to the tail race. This tube of increasing area is known as the Draft Tube. One end of the tube is connected to the outlet of runner while the other end is sub-merged below the level of the water in tail-race.

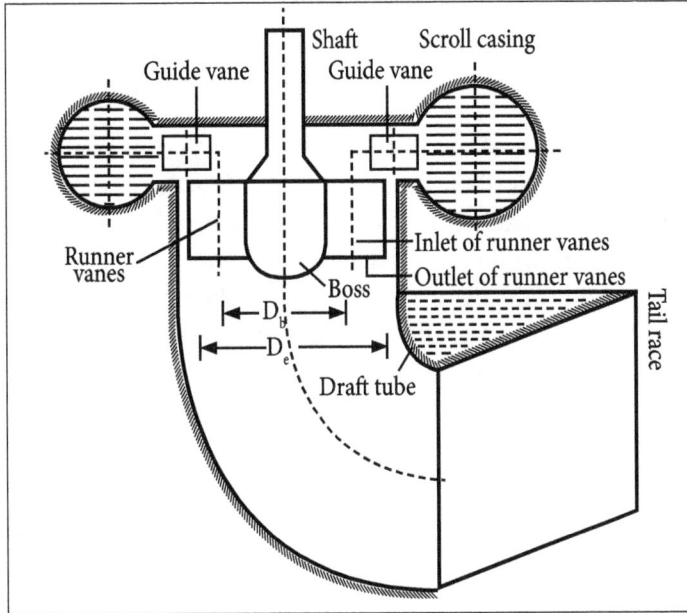

Draft Tube.

Mechanism

The Kaplan turbine is a propeller type turbine and is the most widely used among many variations of the Kaplan Turbine like S, VLH, Propeller, Pit, Bulbs, Straflo and Tyson. The VLH turbine is a very interesting one as it allows the fish to pass through it even during operation with a mortality of fish less than 5%.

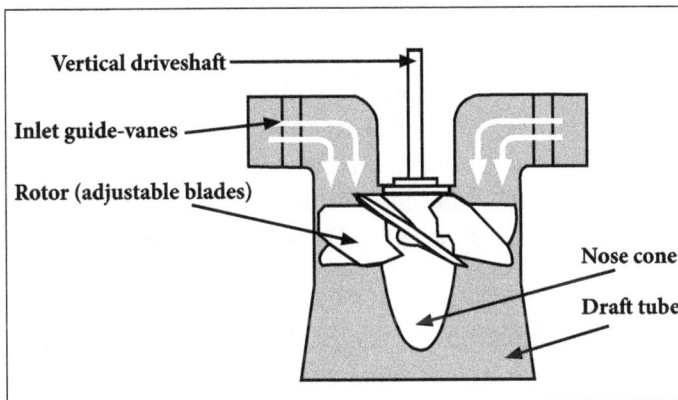

The Kaplan is a simple propeller with adjustable blades and the flow arrive on every side of the rotor as shown on the figure above. The inlet guide vanes can be on position open or close to controlled the amount of water supply in the turbine. The inlet guide vanes also induce the water to turn before the blades to be more efficient. The fact that we can controlled the water entry, enable to have a wide operating discharge and so production of electricity. The nose of the turbine is carefully designed to reduce the losses as long as the geometry of the pipe at the exit to decrease the pressure of water and so all possible energies are available. The blades of the turbine are adjustable as a function of flow velocity, which allow a wide range of discharge, as long as the number of blades.

Even though, the performance of semi-Kaplan's is being compromised when operating across a wide flow range, for applications where the flow does not vary much they are a cost-effective choice. The figure below shows how the efficiency varies across the operating flow range for a full-Kaplan, a semi-Kaplan with adjustable blades and a semi-Kaplan with adjustable inlet guide-vanes. It also shows the efficiency curve for a propeller turbine (a Kaplan with fixed blades and fixed inlet guide-vanes.

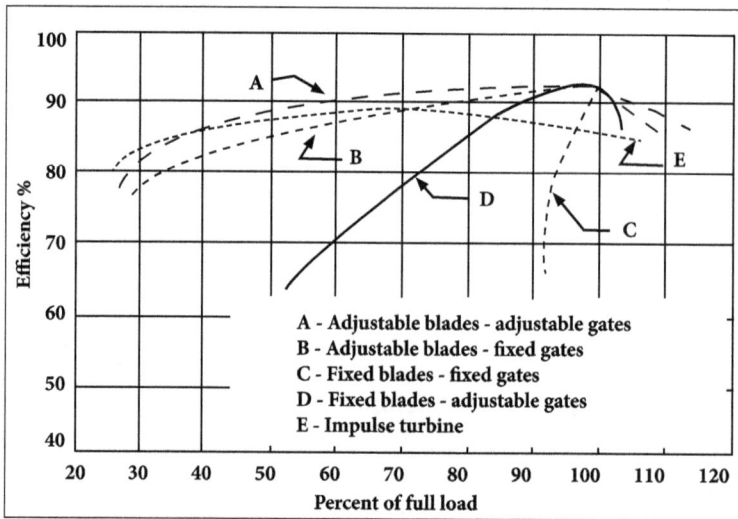

Efficiency of different Kaplan turbine

Kaplan turbines will technically work across a wide range of heads and the flow rates, but as of other turbine types being more effective on higher heads and because Kaplan's are relative expensive, they are the turbine of choice for lower head sites with high flow rates. They are used on sites with net heads from 1.5 to 20 metres and peak flow rates from 3 m³/s to 30 m³/s. In UK, this tends to be on lowland rivers with low heads and relatively high flow rates .These systems will have power outputs ranging from 75 kW up to 1 MW.

The smallest good quality Kaplan turbines available have rotor diameters of 600 mm, though these tend to be prohibitively expensive, atleast a very low heads, so generally speaking the smallest rotors are 800 mm. The largest rotors available have 3 to 5 metre diameters. For even larger sites, multiple-turbines tend to be used rather than increasing the diameter further. Kaplan turbines are available in three basic configurations such as vertical axis, horizontal axis (also referred as S-turbines) and bulb turbines.

Working

The water from the pen stock enters the scroll casing and then moves to the guide vanes. From the guide vanes, the water turns through 90° and flow axially through the runner. The discharge through the runner is obtained as,

$$Q = \frac{\pi}{4}\left(D_o^2 - D_b^2\right) \times V_{f1}$$

Where,

D_o – Outer diameter of the runner.

D_b – Diameter of hub.

V_{f1} – Velocity of flow at inlet.

The inlet and outlet velocity triangles are drawn at the extreme edge of the runner vane corresponding to the points.

Some Important Points for Propeller (Kaplan Turbine)

- The peripheral velocity at inlet and outlet are equal.

$$\therefore u_1 = u_2 = \frac{\pi D_o N}{60}$$

- Velocity of flow at inlet and outlet are equal.

$$\therefore \ V_{f1} = V_{f2}$$

- Area of flow at inlet = Area of flow at outlet.

$$= \frac{\pi}{4}\left(D_o^2 - D_b^2\right).$$

Advantages of Kaplan Turbine

- Very small no of blades are used nearly 3 to 8 blades.

- Less resistance has to be overcome.

- Runner vanes are adjusted in the Kaplan.

- Very low heads are required.

Disadvantages of Kaplan Turbine

- Speed of the turbine is 250 to 850.

- Position of the shaft is only in vertical direction.

- Large Flow rate must be required.

- High speed generator is required.

Problems

1. A Kaplan turbine runner is to be designed to develop 7357.5 kW shaft power. The net available head is 5.50 m. The speed ratio is 2.09, flow ratio is 0.68 and the overall efficiency is 60%. The diameter of the boss is 1/3rd of the diameter of the runner. Let us determine the diameter of the runner, its speed and its specific speed and solver it.

Solution:

Given,

Shaft power, $p = 7357.5$ kw

Head, $H = 5.5$ cm

Speed ratio, $K_u = 2.09$

Flow ratio, $K_f = 0.68$

Formula to be used,

$$K_u = \frac{u_1}{\sqrt{2gH}}$$

$$u_1 = \frac{\pi D_o N}{60}$$

Diameter of boss $(D_b) = 1/3$ diameter of runner (D_o)

Overall efficiency, $\eta_o = 60\%$

$$K_u = \frac{u_1}{\sqrt{2gH}} \Rightarrow u_1 = 2.09 \times \sqrt{2 \times 9.81 \times 5.50}$$

$$u_1 = 21.71 \text{ m/s}$$

$$K_f = \frac{V_{f1}}{\sqrt{2gH}} \Rightarrow V_{f1} = 0.68 \sqrt{2 \times 9.81 \times 5.5}$$

$$V_{f1} = 7.06 \text{ m/s}$$

Overall efficiency, $e_o = \dfrac{\text{Shaft Power } (P)}{\text{Water Power}} = \dfrac{7357.5}{WQH}$

$$Q = \frac{7357.5}{0.60 \times 9.81 \times 5.50}$$

$Q = 227.27 \ m^3/s$

$Q = A \times V_1$

$Q = \dfrac{\pi}{4}\left(D_o^2 - D_b^2\right) \times V_{f1}$

$227.27 = \dfrac{\pi}{4}\left(D_o^2 - 1/9\,D_o^2\right) \times 7.06$

$D_o^2 = 11.57$

$D_o = 3.39\,m$

Speed of the turbine (N),

$u_1 = \dfrac{\pi D_o N}{60}$

$N = \dfrac{60 \times (u_1)}{\pi \times D_o} = \dfrac{60 \times 21.71}{\pi \times 3.39}$

$N = 122.31 \ rpm$

Specific Speed, $N_s = \dfrac{N\sqrt{P}}{H^{5/4}}$

$= \dfrac{122.31 \times \sqrt{7357.5}}{\left(5.50\right)^{5/4}}$

$= \dfrac{10491.24}{8.423}$

$N_s = 1245.58$

2. Let us calculate the diameter and speed of the runner of a Kaplan turbine developing 6000 kW under an effective head of 5 m. Overall efficiency of the turbine is 90%. The diameter of the boss is 0.4 times the external diameter of the runner. The turbine speed ratio is 2.0 and flow ratio 0.6.

Solution:

Given,

Shaft power, S. P. = 6000 kW

Head, H = 5 m

Speed ratio, = 2.0

Flow ratio = 0.6

Overall efficiency, $\eta_o = 90\% = 0.90$

Formula to be used,

$$\eta_o = \frac{S.P}{W.P}$$

$$D_b = 0.4 \times D_o$$

$$Q = \pi / 4 \left(D_o^2 - D_b^2 \right) \times V_{F1}$$

$$u_1 = \frac{\pi D_o N}{60}$$

Speed ratio,

$$u_1 \sqrt{2gh} = 2.0$$

$$u_1 = 2 \times \sqrt{2gH} = 2.0 \times \sqrt{2 \times 9.81 \times 5}$$

$$u_1 = 19.8 \text{ m sec}$$

Flow ratio,

$$\frac{V_{f1}}{\sqrt{2gH}} = 0.6$$

$$V_{f1} = 0.6 \sqrt{2gH}$$

$$= 0.6 \sqrt{2 \times 9.81 \times 5}$$

$$= 5.94 \, \text{m} / \sec$$

$$\eta_o = \frac{S.P}{W.P}$$

$$0.90 = \frac{6000}{\dfrac{\rho \times g \times Q \times H}{1000}}$$

$$0.90 = \frac{6000}{\dfrac{1000 \times 9.81 \times Q \times 5}{1000}}$$

$$Q = \frac{6000 \times 1000}{1000 \times 9.81 \times 5 \times 0.90}$$

$$Q = 136.91 \text{ m}^3/s$$

$$Q = \pi / 4 \left(D_o^2 - D_b^2 \right) \times V_{F1}$$

$$135.91 = \pi/4\left[D_o - (0.4D_o) \times 5.94\right]$$

$$135.91 = 3.92\ D_o^2$$

$$D_o^2 = \frac{135.91}{3.92}$$

$$D_o = \sqrt{\frac{135.91}{3.92}}$$

$$D_o = 5.88\,m$$

$$D_b = 0.4 \times D_o = 0.4 \times 5.88$$

$$D_b = 2.352\ m$$

Speed of the turbine is given by $u_1 = \dfrac{\pi\, D_o\, N}{60}$

$$19.8 = \frac{11 \times 5.88 \times N}{60}$$

$$N = 64.31\ rpm$$

3. A Kaplan turbine runner is to be designed to develop 9100 kW. The net available head is 5.6 m. If the speed ratio = 2.09, flow ratio = 0.68, overall efficiency 86% and the diameter of the boss is 1/3 the diameter of the runner. Let us determine the diameter of the runner, its speed and the specific speed of the turbine.

Solution:

Given,

$$P = 9100\ kw$$

$$H = 5.6\ m$$

Speed ratio = 2.09

Flow ratio = 0.68

Overall efficiency = 86%

Diameter of boss = 1/3 diameter of runner

$$D_b = 1/3\ D_o$$

Formula to be used,

$$\text{Speed Ratio} = \frac{u_1}{\sqrt{2gH}}$$

$$\text{Flow Ratio} = \frac{V_{f1}}{\sqrt{2gH}}$$

$$\eta_o = \frac{P}{\left(\dfrac{\rho g\, QH}{1000}\right)}$$

$$Q = \frac{P \times 1000}{\rho g\, H\, \eta_o}$$

$$\text{Speed Ratio} = \frac{u_1}{\sqrt{2gH}}$$

$$u_1 = 2.09 \times \sqrt{2 \times 9.81 \times 5.6}$$

$$= 21.95 \text{ m/s}$$

$$\text{Flow Ratio} = \frac{V_{f1}}{\sqrt{2gH}}$$

$$V_{f1} = 0.68 \times \sqrt{2 \times 9.81 \times 5.6}$$

$$= 7.12 \text{ m/s}$$

$$\eta_o = \frac{P}{\left(\dfrac{\rho g Q H}{1000}\right)}$$

$$Q = \frac{P \times 1000}{\rho g\, H \eta_o}$$

$$= \frac{9100 \times 1000}{1000 \times 9.81 \times 5.6 \times 0.86}$$

$$= 192.5\, \text{m}^3/\text{s}$$

Discharge through Kaplan turbine,

$$Q = \frac{\pi}{4}\left(D_o^2 - D_b^2\right)V_{f1}$$

$$192.5 = \frac{\pi}{4}\left(D_0^2 - \left(\frac{D_0}{3}\right)^2\right) \times 7.12$$

$$D_0 = \sqrt{\frac{4 \times 192.5 \times 9}{\pi \times 8 \times 7.12}}$$

$$D_0 = 6.12\,m$$

Speed of turbine, $u_1 = \dfrac{\pi DN}{60}$

$$N = \frac{60 \times u_1}{\pi \times D} = \frac{60 \times 21.95}{\pi \times 0.21}$$

$$N = 67.5 \text{ rpm}$$

Specific speed,

$$N_S = \frac{N\sqrt{P}}{H^{5/4}}$$

$$= \frac{67.5\sqrt{9100}}{5.6^{5/4}}$$

$$N_S = 7.46\,\text{rpm}$$

4. A reaction turbine works at 450rpm under a heat of 120 m. Its diameter at inlet is 120 cm and the flow area is 0.4m². The angles made by absolute and relative velocity at inlet are 20° and 60° respectively, with the tangential velocity. Let us determine the volume flow rate, the power developed and the hydraulic efficiency.

Solution:

Given,

Speed of turbine, N = 450rpm

Head, H = 120 m

Diameter at inlet, d_1 = 120cm = 1.2 m

Formula to be used,

$$u_1 = \frac{\pi D_1 N}{60}$$

$$\tan \alpha = \frac{V_{f1}}{V_{w1}}$$

$$\tan\theta = \frac{V_{f_1}}{V_{w_1} - u_1}$$

$$Q = \pi_{D_1} B_1 \times V_{f_1}$$

Work done per second on the turbine $= \rho \cdot Q \left[V_{w1} \, u_1 \right]$

Power developed in KW $= \dfrac{\text{Work done per second}}{1000}$

$$\eta_h = \frac{V_{w_1} u_1}{g H}$$

Flow Area, $\pi d_1 \times B_1 = 0.4 m^2$

Angle made by absolute velocity at inlet, $\alpha = 20°$

Angle made by the relative velocity at inlet, $G = 60°$

Whirl at outlet, $V_{w2} = 0$

Tangential velocity of the turbine at inlet,

$$u_1 = \frac{\pi D_1 N}{60} = \frac{\pi \times 1.2 \times 4.50}{60} = 28.27 \text{ m/s.}$$

From inlet velocity triangle,

$$\tan \alpha = \frac{V_{f1}}{V_{w1}} \text{ or } \tan 20° = \frac{V_{f1}}{V_{w1}} = \tan 20° = 0.364$$

$$\therefore V_{f_1} = 0.364 \, V_{w1} \qquad \qquad \dots(1)$$

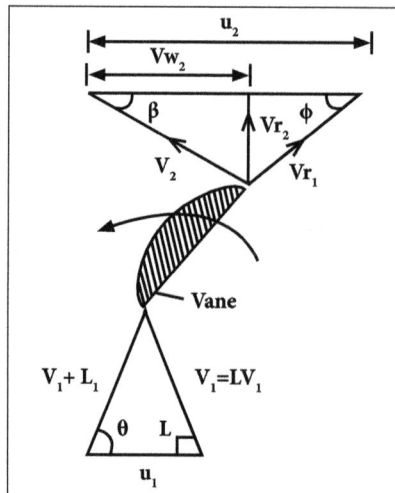

Also,

$$\tan\theta = \frac{V_{f_1}}{V_{w_1} - u_1} = \frac{0.364\,V_{w_1}}{V_{w_1} - 28.27} \quad \left(\because V_{f_1} = 0.364\,V_{w_1}\right)$$

or,

$$\frac{0.364\,V_{w_1}}{V_{w_1} - 28.27} = \tan\theta = \tan 60° = 1.732$$

$$\therefore\ 0.364\,V_{w_1} = 1.732\left(V_{w_1} - 28.27\right) = 1.732\,V_{w_1} - 48.96$$

or,

$$\left(1.732 - 0.364\right) V = 48.96$$

$$\therefore\ V_{w_1} = \frac{48.96}{\left(1.732 - 0.364\right)} = 35.789 = 35.79 \text{ m/s}$$

From equation (1),

$$V_{f_1} = 0.364 \times V_{w_1} = 0.364 \times 55.79 = 13.027 \text{ m/s}$$

i. Value flow rate is given by,

$$Q = \pi_{D_1}\,B_1 \times V_{f_1}$$

But,

$$\pi d_1 \times B_1 = 0.4 \text{ m}^2;$$

$$Q = 0.4 \times 13.027 = 5.211 \text{ m}^3/\text{s}$$

ii. Work done per second on the turbine is given by,

$$= \rho \cdot Q\left[V_{w1}\,u_1\right]\left(\because V_{W2} = 2\right)$$

$$= 1000 = 5.211\left[35.79 \times 28.27\right] = 5272402 \text{ Nm/s}$$

$\therefore$ Power developed in KW

$$= \frac{\text{Work done per second}}{1000} = \frac{5272402}{1000} = 5272.402 \text{ KW}$$

iii. The hydraulic efficiency is given by,

$$\eta_h = \frac{V_{w_1} u_1}{gH} = \frac{35.79 \times 28.27}{9.81 \times 120} = 0.8595 = 85.95\,\%$$

5.3   Governing of Turbines

Hydraulic turbines are directly coupled to the electric generators. The generators are always required to run at a constant speed irrespective of the variations in the load. This constant speed (rpm) of the generator is given by,

$$N = \frac{120f}{p}$$

Where f is the frequency for power generated in cycles per second and p is the number of poles for the generator. The speed of the generator can be maintained at a constant level only if the speed of the turbine runner is constant as given by Equation.

It is known as the synchronous speed of the turbine runner for which it is designed. If the load on the generator goes on varying and if the input for the turbine remains the same, then the speed of the runner tends to increase if the load goes down or it tends to decrease if the load on the generator goes up.

Therefore, the speed of the generator and hence, the frequency will vary according-ly, which is not desired. Therefore, the speed of the runner is always required to be maintained at a constant level at all loads. It is done automatically by a governor which regulates the quantity of water flowing through the runner in proportion to the load.

Governing of Impulse Turbine

In a Pelton turbine, water flow to the runner is regulated by the combined action of the spear and the deflector plate. There is a centrifugal governor, as in the case of a steam turbine, where its sensitivity to load variation is augmented by an oil-operated servo-mechanism.

When the load on the generator drops, the speed of turbine runner increases. The fly balls of the centrifugal governor fly outward due to more centrifugal force (due to high-er rpm). The sleeve moves up, the portion of the lever to the right of the fulcrum moves down pushing the piston rod of the control valve downwards.

With the downward motion of the piston rod valve V_1 closes and valve V_2 opens as shown in Figure. A gear pump pumps oil from the oil sump to the control or relay valve. Oil flows through valve V_2 and exerts force on the face L of the piston of the relay cylinder. The piston (or spear) rod along with the spear moves to the right, thus decreasing the flow area and hence, the rate of water flow to the turbine.

The speed of the turbine falls till it becomes normal when the fly balls, sleeve, lever, etc. also come to normal position. The reverse happens when the load on the generator increases, speed decreases, fly balls fly inward with less centrifugal force (due to less rpm), the sleeve moves down, the piston rod of control valve goes up, valve V_1 opens and valve V_2 closes, the oil under pressure flows through valve V_1 and exerts a force on the face M of the piston. The piston rod and the spear move to the left as a result of which more water flows to the turbine to take up more load and the speed becomes normal, i.e. attains its rated value.

The spear or needle valve is used normally for small load fluctuations. When there is a sudden fall of load, the spear has to move rapidly to close the nozzle. This rapid closing may cause water hammer. It is quite serious in large capacity plants with long penstocks. To avoid the water hammer effects during a sudden fall of load, a deflector is introduced in the system, which is not shown in Figure. The function of the deflector is to deflect some water from the jet advancing to the turbine runner when the load on the turbine suddenly decreases. The quantity of water flowing through the nozzle remains the same, but a certain part of water coming out from the nozzle is deflected and is not allowed to strike the buckets. The deflected water goes waste into the tailrace level.

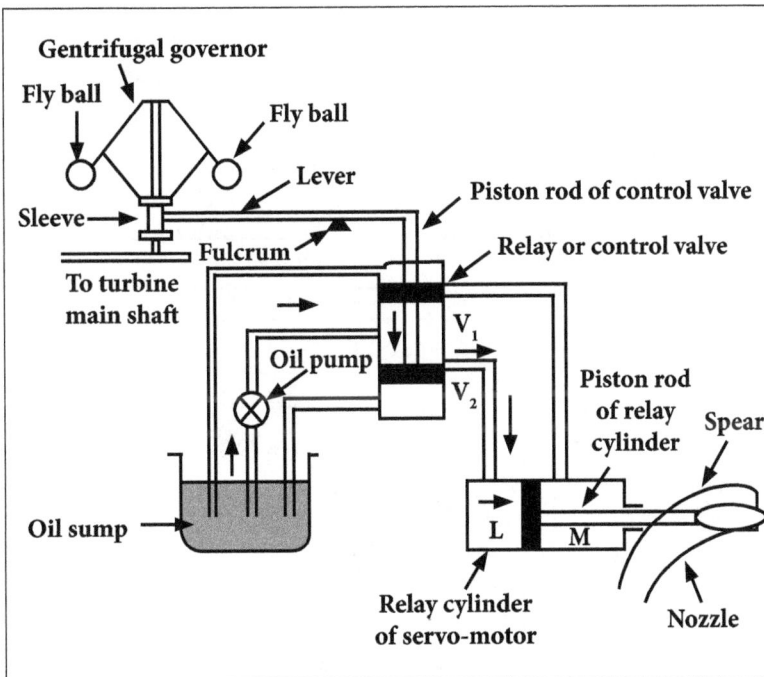

Governing of Pelton turbine.

5.3.1 Performance and Characteristic Curves

Characteristic Curves

The following are the important characteristic curves of a turbine:

- Main characteristic curve.
- Operating characteristic curve.
- Constant efficiency curve.

Main Characteristic Curves

Main characteristic curves are obtained by maintaining a constant head and a constant gate opening (G.O) on the turbine. The speed of the turbine is varied by changing load on the turbine. For each value of the speed, the corresponding value of power (P) and discharge (Q) are obtained. Then the overall efficiency (η_0) for each value of the speed is calculated.

From these readings the value of unit speed (NU) unit power (PU) and unit discharge (QU) are determined. Taking Nu as abscissa, the values of Qu, Pu, P and η_0 are plotted. By changing the gate opening, the values of Qu, Pu and η_0 and Nu are determined are taking Nu as abscissa the value of Qu, Pu and η_0 are plotted.

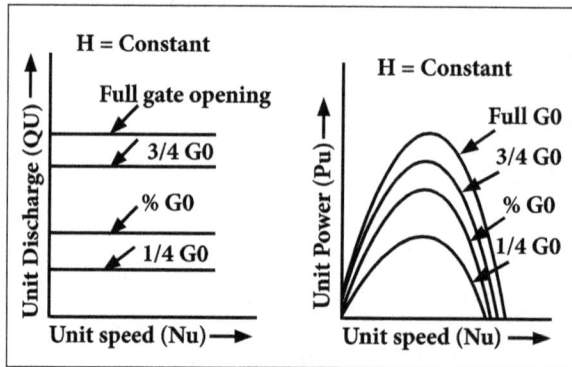

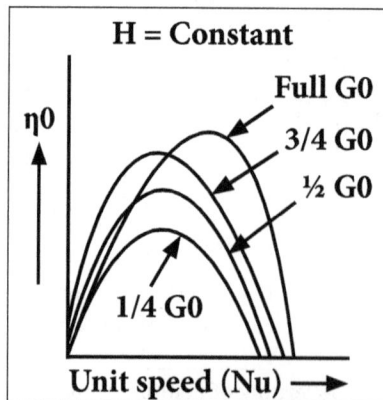

Main characteristic curve.

Operating Characteristic Curves

Operating characteristic curves are plotted when the speed on turbine is constant. In case of turbines, the head is generally constant. There are three independent parameters namely N, H and Q. For operating characteristics N and H are constant and hence the variation of power and efficiency with respect to discharge Q are plotted. The power and efficiency curves will be slightly away from origin on x-axis to overcome initial friction certain amount of discharge will be required.

Operating characteristic curves are plotted when the speed on turbine is constant. In case of turbines, the head is generally constant. There are three independent parameters namely N, H and Q. For operating characteristics N and H are constant and hence the variation of power and efficiency with respect to discharge Q are plotted. The power and efficiency curves will be slightly away from origin on x-axis to overcome initial friction certain amount of discharge will be required.

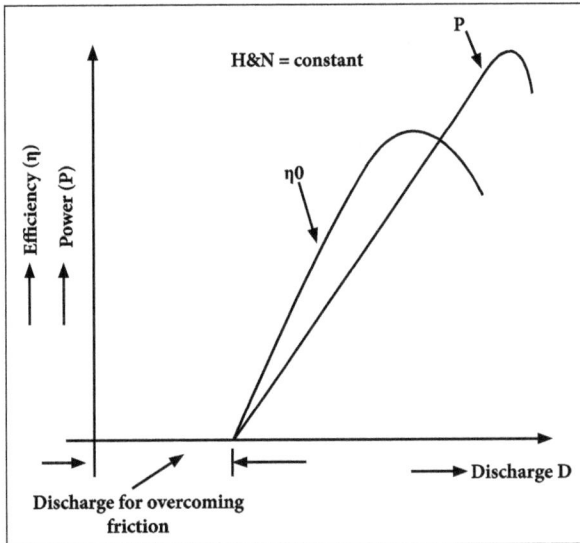

Operating characteristic curve.

These current are obtained from speed Vs efficiency and speed Vs discharge curves for different gate opening. For a given efficiency from Nu Vs ηo curves, there are two speeds. From Nu Vs Qu curve, corresponding to two values of speeds there are two values of discharge. This means for a given efficiency, there are two values of speeds and two values of discharge for a given gate opening.

If the efficiency is maximum, it has only one value. These two values of speed and two values of discharge corresponding to a particular gate opening are plotted. The procedure is repeated for different gate opening and the curve Q Vs N are plotted.

The points having the same efficiency are joined. The curves having same efficiency are called is co-efficiency curves. These curves are helpful for determining the zero of constant efficiency and for predicating the performance of turbine at various efficiencies.

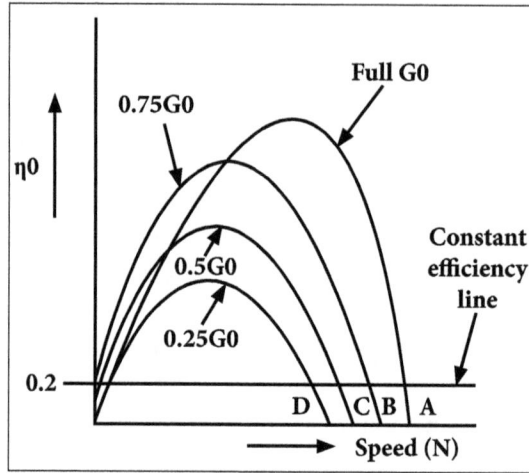

Constant Efficiency Curve.

Hydro Power

6.1 Hydro Power: Components of a Hydroelectric Power Plant

The power system mainly consists of three parts namely generation, transmission and distribution. Generation refers how to generate the electricity from the available source. There are several methods to generate electricity but here we focused only on generation of electricity by means of hydro or water.

In hydro power plant we here use gravitational force of fluid water to run the turbine which is coupled with the electric generator in order to produce electricity. It plays a main role to protect our fossil fuel that is limited, because the generated electricity in hydro power station by the use of water which is the renewable source of energy and it is available in a large amount without any cost.

The biggest advantage of hydro power is the water which the main stuff to produce electricity in hydro power plant, is free and does not contain any type of pollution. And after generating electricity the price of electricity is average and not too much high.

Essential Elements

1. Dam

The dam is the most important component of the hydroelectric power plant. The dam is built upon a large river that has abundant quantity of water throughout the year. It must be built at a location where the height of the river is sufficient to get maximum possible potential energy from the water.

2. Water Reservoir

The water reservoir is a place behind the dam where water is being stored. The water in the reservoir is located higher than rest of the dam structure. The height of water in the reservoir shows how much potential energy does the water possesses. The higher the height of the water, the more its potential energy. The high position of water in the reservoir also enables the water to move downwards effortlessly.

The height of the water in reservoir is higher than natural height of water flowing in the river, and so it is considered to have an altered equilibrium. This also helps in increasing the overall potential energy of water, that in turn helps ultimately in producing more electricity in the power generation unit.

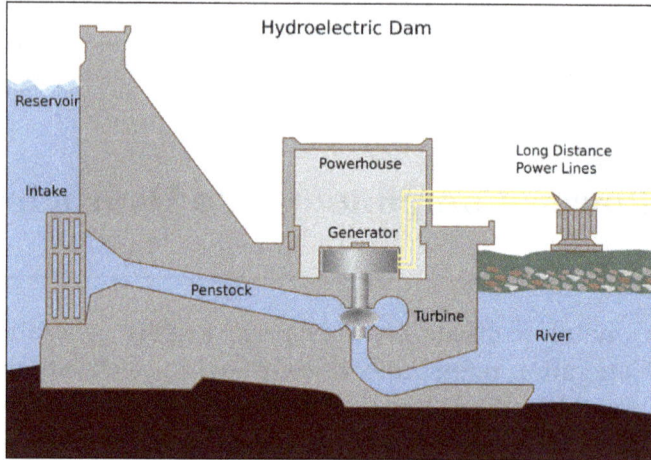

Hydel Power plant- Essential elements.

3. Intake or Control Gates

It is built inside the dam. The water from the reservoir is released and they are controlled through these gates. These are known as inlet gates because water enters the power generation unit with the help of these gates. When the control gates are opened, the water flows due to gravitational force through the penstock and also towards the turbines. The water flowing through the gates possesses potential as well as the kinetic energy.

4. The Penstock

It is a long pipe or a shaft which carries the water flowing from the reservoir towards power generation unit, that comprises of the turbines and generators. The water in the penstock possesses the kinetic energy due to its motion and potential energy due to its height.

The total amount of power generated in the hydroelectric power plant depends upon the height of the water reservoir and also the amount of water flowing through the penstock. The amount of water flowing through the penstock is controlled by control gates.

5. Water Turbines

Water flowing from the penstock is allowed to enter the power generation unit that houses the turbine and the generator. When the water falls on the blades of turbine the kinetic and the potential energy of the water is converted into the rotational motion of the blades of the turbine. The rotating blades causes the shaft of the turbine to also rotate. The turbine shaft is enclosed inside the generator. In most hydroelectric power plants there are more than one power generation unit.

There is a great difference in the height between the level of the turbine and the level of water in reservoir. This difference in height is also called as the head of water that decides the total amount of power that will be generated in the hydroelectric power plant.

There are several types of water turbines such as Kaplan turbine, Francis turbine, Pelton wheels etc. The type of turbine used in the hydroelectric power plant depends upon the height of the reservoir, quantity of water and total power generation capacity.

6. Generators

It is in the generator where the electricity is produced. The shaft of the water turbine rotates in a generator produces alternating current in the coils of the generator. It is the rotation of the shaft inside the generator which produces the magnetic field that is converted into electricity by means of electromagnetic field induction.

Therefore, the rotation of the shaft of the turbine is crucial for the production of electricity and this is achieved by means of the kinetic and potential energy of the water. Hence in hydroelectricity power plants potential energy of water is being converted into electricity.

6.1.1 Pumped Storage Systems

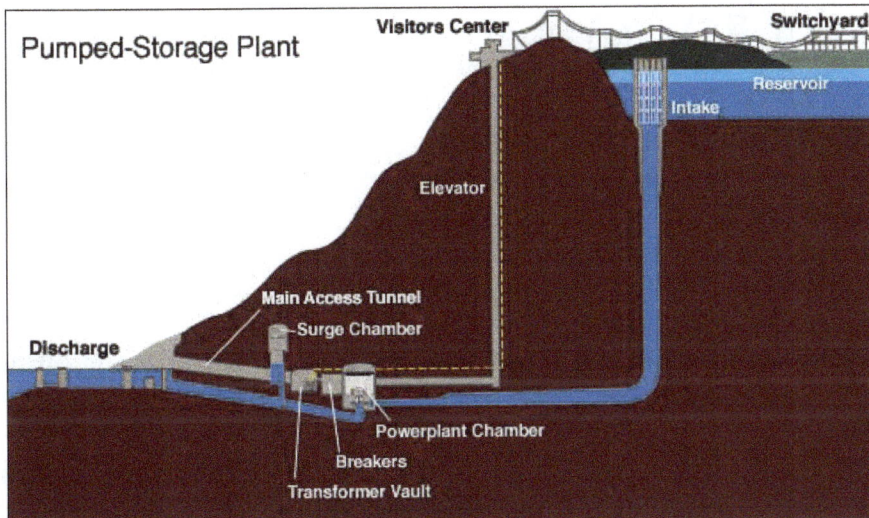

Pumped storage systems.

Pumped storage is an essential solution for the grid reliability, providing one of the few large-scale, reasonable means of storing and deploying the electricity. The Pumped storage projects store and generate the energy with the help of moving water between the two reservoirs at different elevations. At times of low electricity demand, such as at night or on weekends, excess energy is used in order to pump water to an upper reservoir. During periods of high electricity demand, the stored water is released through the turbines in the same manner as a conventional hydro station, flowing down the hill from the upper reservoir into the lower and then generating electricity. The turbine is then able to also act as a pump, moving water back up the hill.

Pumped storage is one of the most cost-effective utility-scale options for grid energy storage, acting as a key provider of what is termed as the ancillary services. Ancillary services includes network frequency control and reserve generation ways of balancing electricity across a large grid system. With an ability to respond almost instantaneously to changes in the amount of electricity running through grid, pumped storage is an essential component of nation's electricity network.

6.1.2 Estimation of Water Power Potential

Electricity from water is basically referred to as Hydro-Power, where the term 'hydro' is the Greek word for water and hydropower is energy contained in the water. It may be converted in the form of electricity through the hydroelectric power plants. All that is required is a continuous inflow of water and difference of height between the water level of the upstream intake of the power plant and its downstream outlet. In order to evaluate the power of flowing water, we may assume a uniform steady flow between the two cross-sections of a river, with H (metres) of difference in the water surface elevation between the two sections for a flow of Q (m³/s), the power (P) is expressed as,

$$P = \gamma Q \left(H + \frac{v_1^2 - v_2^2}{2g} \right) [\text{Nm/s}]$$

Where v_1 and v_2 are the mean velocities in two sections. Neglecting slight differences in the kinetic energy and assuming a value of γ as 9810N/m², one obtains the expression of power as,

$$P = 9810 QH \quad [\text{Nm/s}]$$

Since an energy of 1000Nm/s is represented as 1kW (1kilo-Watt), one can write the following,

$$P = 9.81 \, QH \quad [\text{kW}]$$

The above expression gives a theoretical power of the selected river stretch at a specified discharge. In order to evaluate the potential of the power that may be generated by harnessing the drop in the water levels in a river between the two points, it is needed to have knowledge of the hydrology or stream flow of the site, as that would be varying every day. Even the average monthly discharges over a year would vary. Similarly, these monthly averages will not be the same for consecutive years.

Hence, in order to evaluate the hydropower potential of a site, the following criteria are considered:

- Minimum potential power is based on the smallest runoff available in the stream at all times, days, months and even years having duration of 100 percent. This value is generally of small interest.

- Small potential power is calculated from the 95 % duration discharge.

- Medium or average potential power is gained from the 50 % duration discharge.

- The Mean potential power results by means of evaluating the annual mean run-off. As it is not economically feasible to harness the entire runoff of the river during flood, there is no reason for including the entire magnitude of the peak flows while calculating potential power or potential annual energy.

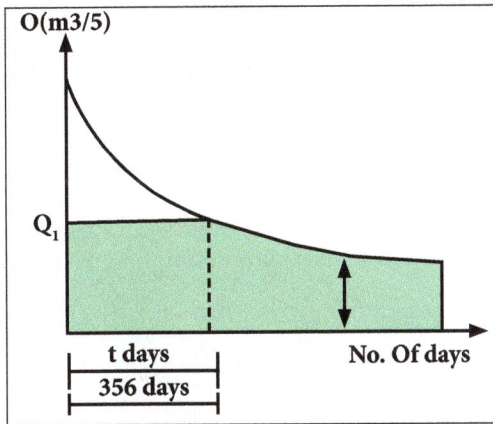

Expressed in time.

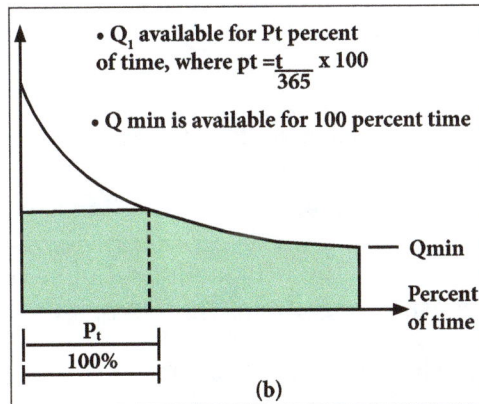

Expressed in percentage.

Therefore, a discharge-duration curve is prepared to figure which plots the daily discharges at a location in decreasing order of magnitude starting from largest daily discharge observed during the year and going up to minimum daily discharge.

From this annual discharge curve, a truncation is being made at a discharge Qt which is the discharge corresponding to a time of 't' days, here t can be the median (say, 182 days or 50 percent duration, denoted by (Q182 or Q50%), or a higher Qt (t less than 182 days) may be selected by specialists who are familiar with local conditions and future plans for power supply. Accordingly, the annual magnitude of potential energy is computed in KWh as below and referring to the above figure,

$$E_p = 24 \times 9.81\, H \left(Q_1 + \sum_i^{365} Q_i \right)$$

$$\approx 235\, H.A \quad (\text{in kWh})$$

Where Qi denotes the daily mean flow during the period 365-t days and A, the hatched area cut by Q_t, where the area under the curve has a unit m³ × day/s.

6.2 Estimation of Load on the Turbines

Significance of the Load Curve

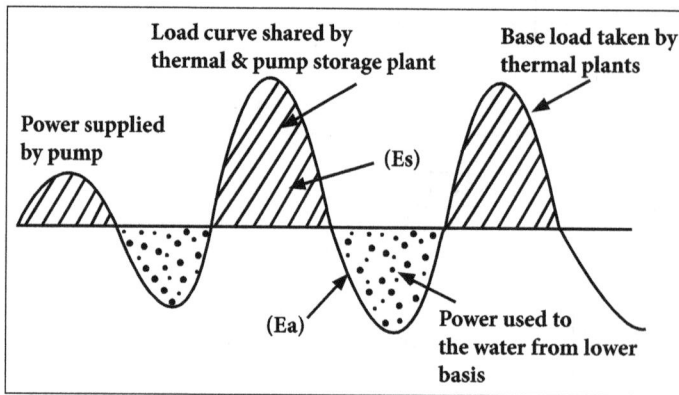

Significance of Load Curve.

Importance of the Load Curves

The load curve is used to represent the rearrangement of all the load elements of load curve in order of their decreasing magnitude. This curve is derived from the load curve.

The load curves supplies the following information:

- The variation of the load during different hours of the day.
- The area under the curve represents the total number of units generated in a day.
- The ratio of the area under the load curve to the total area of the rectangle in which it is contained gives the load factor.

- The peak of curve which represents the maximum demand on the station on the particular day.

- The area under the load curve divided by the number of hours represents the average load on the power station.

Types of Load Curves:

- Yearly load curve - Load curve obtained from the monthly load curve.

- Monthly load curve – Load curve obtained from the daily load curve.

- Daily load curve – Load variations during the whole day.

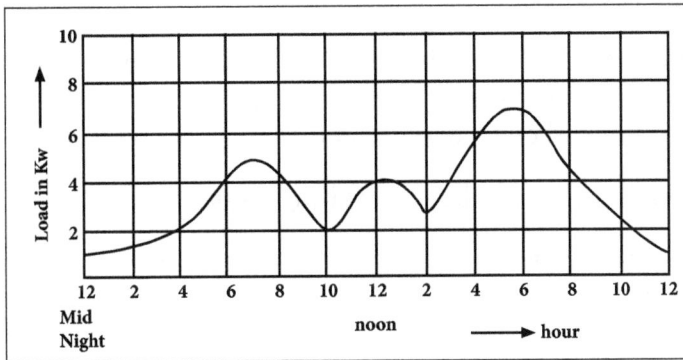

Load curve.

Load Characteristics

- Maximum demand.
- Plant use factor.
- Load factor.
- Connected load.
- Average load.
- Diversity factor.
- Plant capacity factor.

Capacity Factor

The capacity factor is the ratio of the average load on the machine for a period of time considered, to the rating of the machine.

Utilization Factor

In normal operating conditions, the power consumption of a load is sometimes less than indicated, as its nominal power rating, a fairly common occurrence that justifies the application of utilization factor (ku) in the estimation of realistic values.

Utilization Factor = The time that an equipment is in use. In other words, The total time that it may be in use.

Example: The motor can only be used for eight hours a day, 50 weeks a year. The hours of operation would then be 2000 hours, and the motor Utilization factor for a base of 8760 hours per year would be 2000/8760 = 22.83%. With a base of 2000 hours per year, the motor Utilization factor will be 100%.

The bottom line is that the use factor is applied in order to get the correct number of hours that the motor is in use.

This factor should be applied to each individual load, with particular attention to the electric motors that are very rarely operated at full load. In an industrial installation this factor will be estimated on an average at 0.75 for motors.

For incandescent-lighting loads, the factor always equals to 1.

For socket-outlet circuits, the factors depend entirely on the type of appliances being supplied from the sockets concerned

Problems

1. Let us consider that a power generating station has a maximum demand of 100 kW and the daily load on the station is as follows:

Period	kw
6 am to 8 am	3500
8 am to 12 noon	8000
12 noon to 1 pm	3000
1 pm to 5 pm	7500
5 pm to 7 pm	8500
7 pm to 9 pm	1000
9 pm to 11 pm	4500
11pm to 6 am	2000

Here we will learn to draw the load curve and load duration curve and also to Calculate the load factor, plant capacity tested and plant use factor of the power station.

Solution:

The load curve and the load duration curve can be drawn as follows:

Period load curve

Period load duration curve

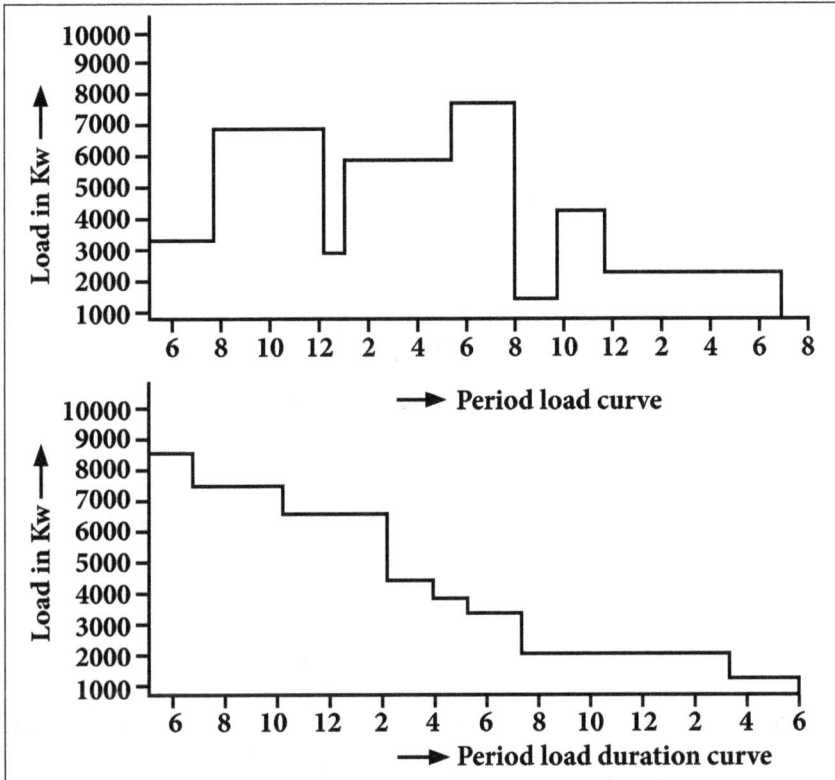

The load curve and load duration curve are drawn as shown in figure 1 and figure 2 respectively.

Energy generated = Area under load curve,

$$= (13500 \times 2) + (8000 \times 4) + (3000 \times 1) + (7500 \times 4) + (8500 \times 2) + (1000 \times 2)$$
$$+ (4500 \times 2) + (2000 \times 7)$$

$$= 7000 + 32000 + 3000 + 30000 + 17000 + 2000 + 9000 + 14000$$
$$= 114000 \text{ Kwh}$$

Average load $- \dfrac{114000}{24} = 4750 \text{ kw}$

Maximum load = 8500 kW,

Load factor $= \dfrac{\text{Average load}}{\text{Maximum load}}$

$$= \dfrac{4750}{8500} = 0.55$$

$$\text{Plant capacity factor} = \frac{\text{Energy generated}}{\text{Capacity of the plant} \times \text{Operating}}$$

$$= \frac{114000}{1000 \times 24}$$

$$= \frac{114000}{24000} = 4.75$$

$$\text{Utilisation factor} = \frac{\text{Maximum load}}{\text{Rate capacity of the plant}}$$

$$= \frac{8500}{1400} = 8.5 \text{ kw}$$

2. Let's we see an example for the maximum demand of a power station is 80000 KW and daily load is described as follows:

Time (hrs)	0-6	6-8	8-12	12-14	14-18	18-22	22-24
Load (M W)	40	50	60	50	70	80	40

Here we determine the load factor of power station and draw the Load duration curve and also lets find the load factor of standby equipment rated at 25 MW that takes up all load in excess of 60 MW.

Solution:

Load Curve Graphs:

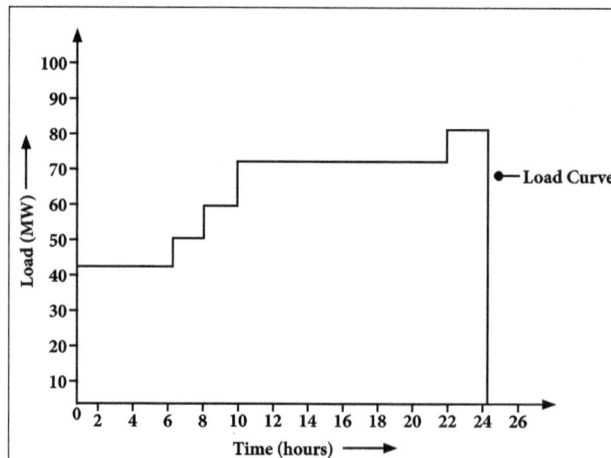

Energy generated $= 40 \times 6 + 50 \times 2 + 60 \times 4 + 50 \times 2 + 70 \times 4 + 80 \times 4 + 40 \times 2 = 1360 \text{MWh}$

Average Load $= \dfrac{1360}{24} = 56.667 \text{ MW}$

$$\text{Load Factor} = \frac{\text{Avg. Load}}{\text{Max. Demand}} = \frac{56.667}{80} = 0.708$$

6.2.1 Diversity Factor: Load and Duration Curve

A consumer uses the power supplied to him at his discretion. If he uses the whole of the connected load, his load factor is 100 per cent. If the maximum used load is less than the connected load which is usually the case, his load factor is lower than 100 per cent. Each consumer uses power for a variety of purposes and touches his maximum demand at certain times. And not necessarily do the maximum demands of all consumers coincide at a particular time However; there is a period during which the combined consumption of all the consumers is maximum. The ratio of the sum of the maximum demands of the individual consumers and the simultaneous maximum demand of all the consumers during a particular time is known as 'diversity factor'. Thus:

Diversity factor = Sum of individual maximum demands / Simultaneous maximum demand.

Since the usage pattern of e consumers is diverse, the numerator in the above expression is greater than the denominator with the result that the diversity factor is more than unity. A higher factor improves the overall load factor because of less maximum demand, and with low installed capacity, more energy could be supplied on account of greater diversity, thereby reducing the capital cost of the station.

Load Duration Curve

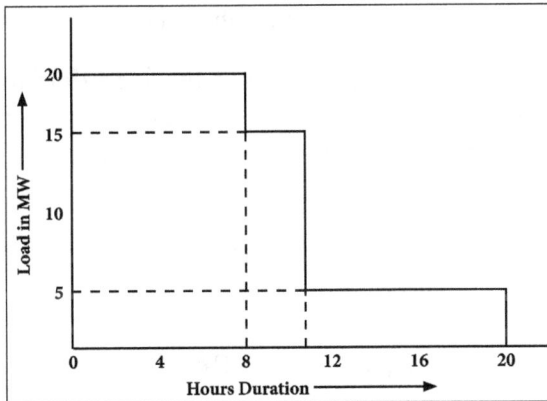

Load duration curve.

Load duration curve is the plot of load in kilowatts versus time duration for which it occurs. When the elements of a load curve are arranged in the order of descending magnitudes.

The load curve is obtained from the same data as the load curve but the ordinates are arranged in the order of descending magnitudes. In other words, the maximum load is represented to the left and decreasing loads are represented to the right in the descending order. Hence the area under the load duration curve and the load curve are equal.

Integrated Load Duration Curves

A graph of a number of units generated (kWh) for a given demand (kW) is called integration load duration curve on Y-axis, load demand in kW or MW is plotted while on z-axis corresponding number of units generated are obtained. Such a curve corresponding to load duration curve shown in figure below.

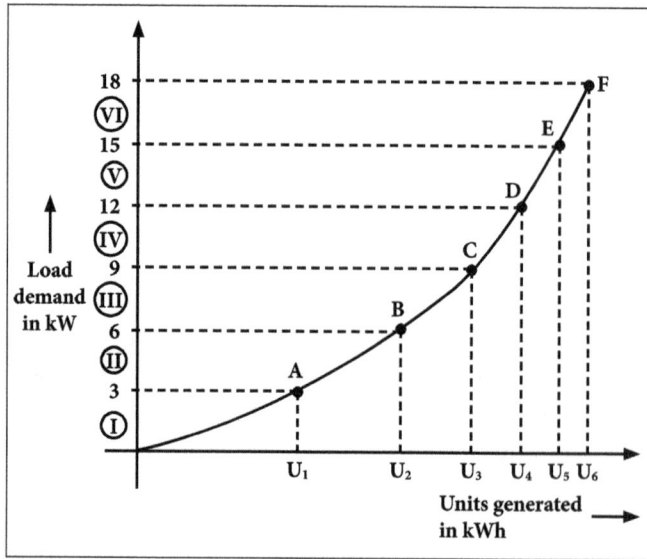

Integrated Load Duration Curve.

This curve is got from load duration curve. Let the load demand be 3 kW from the load duration curve in section I. The number of units produced corresponding to this demand will be area under section I which is shown as U_1 in integrated load duration curve. Similarly the other pints are also obtained to get a total curve.

The number of units consumed by a load up to a particular time of a day can also be shown on a curve which is called as mass curve.

6.2.2 Firm Power, Secondary Power, Prediction of Load

Firm Power

Firm or primary, power is the power that is always ensured to a consumer at any hour of the day and is, thus, completely dependable power. Such a power would correspond to the minimum stream flow and is available for all times. It could, however, be increased by using pondage and as such, does not necessarily correspond to the continuous 24-hour flow available 100 per cent of the lime. Thus firm capacity depends on the minimum stream discharge at the time of peak load, pondage available, shape and size of the connected load curve and the interconnection of other existing plants.

A hydro-electric plant commonly has a wheel capacity corresponding to the flow available for about 20-40% of the time. This percentage depends upon the load factor, the load conditions and the place of the hydel plant in the inter-connected system. Justin and Creaser define firm capacity of a hydel plant as that portion of its total installed capacity which can perform the same function on that portion of the load curve assigned to it as alternative stream capacity could perform.

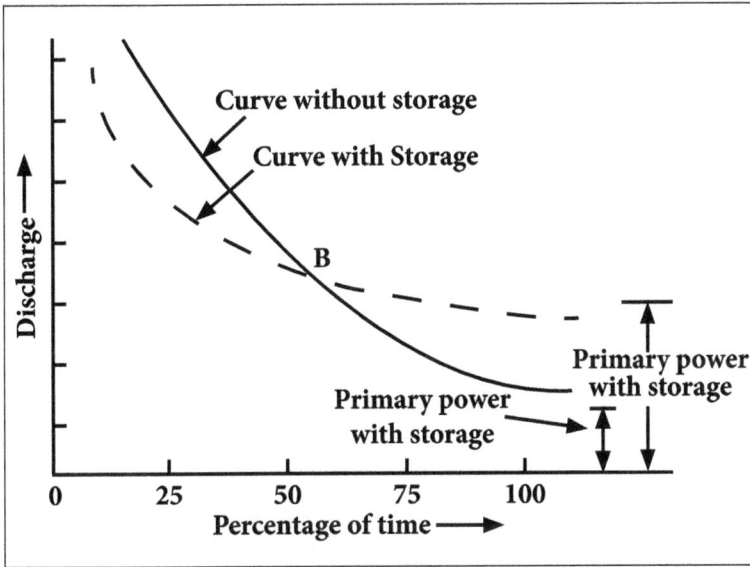

Discharge regulation with storage.

A natural flow-duration curve can be modified with the help of storage regulation. From Figure, it will be noticed that the curve flattens so that it is higher than the original curve on the downstream side of point B when the storage regulation is done. Thus, a greater minimum flow is available for 100% of the time. At any instant, the power is proportional to discharge and can be estimated as firm power.

Secondary Power

Also known as surplus or non-fum, power secondary power is comparatively less valuable. In Figure the feasible primary power and secondary power are shown with reference to the flow-duration curve. If the power is available intermittently at unpredictable times, then also it is called secondary power.

Secondary power is useful in an inter-connected system of power plants. At off-peak hours, secondary power may be called upon to relieve the inter-connected stations, affecting the economy. Secondary power may also be used to take care of the current demand by following a load-sharing plan.

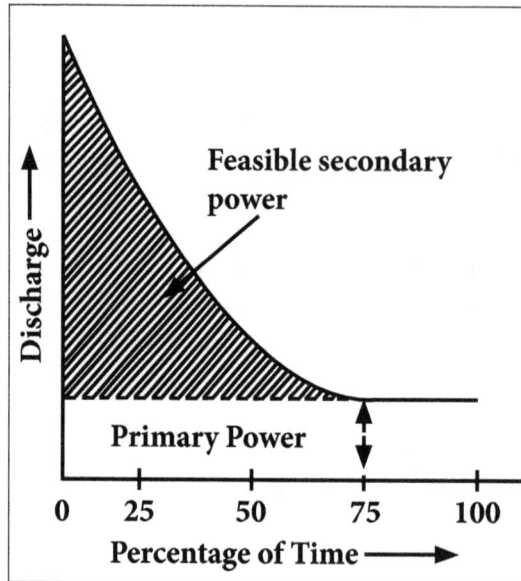

Primary and secondary power.

Prediction of Load

Load prediction or forecasting may be required for:

- Short-term, covering a period of 4 to 7 years.

- Medium-term, covering a period of 8 to 15 years.

- Long-term, covering a period of about 15 years or more.

Whereas the short-term forecasting is done for the areas of deficit or surplus power for operation-planning, medium-term forecast is the basis for the expansion programme of power generation transmission facilities. Long-term forecast helps in the formulation of a country's perspective plan for power generation, vis-a-vis its resources and modes of transmission of voltages. Many methods of forecasting the load demand are available which could be used depending upon the degree of accuracy required and circumstances prevailing. Some of these methods take into consideration the following factors:

- Class-wise consumption.

- Mathematical formulae.

- Historical trends.

- Correlation between power development and economic development.

In addition to these methods, there are a large number of formulae deploying simple extrapolation technique from past records, to using complex correlations. One of such formulae, for instance, is the Scheer formula for estimating generation requirement, as given below,

$$\log_{10} G = c - 0.15 \log_{10} U$$

$$G = \frac{10^c}{U^{0.15}}$$

Where,

G - is the annual growth in generation (per cent).

U - The per capita generation.

C - A constant = 0.02(population growth rate) + 1.33.

The constant 0.15 (which is 1/2log 2) has been derived on the basis of thumb rule that 'a hundred-fold increase in per capita generation will reduce the rate of growth by half.

For prediction of generation, one should start with a year for which the generation is known, along with the starting year's population and forecast of population of future years. Thus, the value of C and per capita generation for the starting year are worked out and then the rate of growth G is evaluated by the above formula. Generation in the next year is calculated by multiplying the generation in the starting year by (1+G)/100.

Thus, generation for the next year and forecast for population being known, the values of C and U for the next year are evaluated and G for subsequent year can be calculated. In this way, the year-by-year prediction is done.

Another formula for the prediction of load is the Belgium formula, given below,

$$E = KM^{0.6} \left(2\right)^{0.465t}$$

Where,

E- is the electricity consumption.

M- is the index of manufacture or production.

T - is the time for which consumption is to be projected.

K- is adjustment factor.

Some countries have derived specific formulae to suit their economic situation and growth trend, just as the one given above used in Belgium.

Problems

1. A diesel station supplies the following loads to various consumers.

Industrial consumer = 1500 kW; Commercial establishment = 750 kW.

Domestic power = 100 kW; Domestic light = 450 kW.

If the maximum demand on the station is 2500 kW and the number of kWh generated per year is 45 × 105, let us determine (i) the diversity factor and (ii) annual load factor.

Solution:

i. Diversity factor $= \dfrac{1500 + 750 + 100 + 450}{2500} = 1.12$

ii. Average demand $= \dfrac{\text{k Wh generated / annum}}{\text{Hours in a year}} = 45 \times 10^5 / 8760 = 513.7$ kW

$\therefore$ Load factor $= \dfrac{\text{Average load}}{\text{Max. demand}} = \dfrac{513.7}{2500} = 0.205 = 20.5\%$

2. A power station has a maximum demand of 15000 kW. The annual load factor is 50% and plant capacity factor is 40%. Let us determine the reserve capacity of the plant.

Solution:

Energy generated/annum = Max. Demand × L.F. × Hours in a year

$$= (15000) \times (0.5) \times (8760) \text{ kWh}$$

$$= 65.7 \times 106 \text{ kWh}$$

Plant capacity factor $= \dfrac{\text{Units generated / annum}}{\text{Plant capacity} \times \text{Hours in a year}}$

$\therefore$ Plant capacity $= \dfrac{65.7 \times 10^6}{0.4 \times 8760} = 18{,}750$ kW

Re serve capacity = Plant capacity − Max. demand

$$= 18{,}750 - 15000 = 3750 \text{ k W}$$

3. A power supply is having the following loads:

Type of Load	Max. demand (kW)	Diversity of group	Demand factor
Domestic	1500	1.2	0.8
Commercial	2000	1.1	0.9
Industrial	10,000	1.25	1

If the overall system diversity factor is $1 \cdot 35$, let us determine (i) the maximum demand and (ii) connected load of each type.

Solution:

i. The sum of maximum demands of three types of loads is $= 1500 + 2000 + 10,000 = 13,500$ kW.

As the system diversity factor is $1 \cdot 35$,

$\therefore$ Max. Demand on supply system $= 13,500 / 1 \cdot 35 = 10,000$ kW

ii. Each type of load has its own diversity factor among its consumers.

Sum of max. demands of different domestic consumers,

$\qquad$ = Max. Domestic demand $\times$ diversity factor

$\qquad = 1500 \times 1 \cdot 2 = 1800$ kW

$\therefore$ Connected domestic load $= 18000 / 8 = 2250$ kW

Connected commercial load $= 2000 \times 1 \cdot 10 \cdot 9 = 2444$ kW

Connected industrial load $= 10,000 \times 1 \cdot 251 = 12,500$ kW

4. At the end of a power distribution system, a certain feeder supplies three distribution transformers, each one supplying a group of customers whose connected loads are as under:

Transformer	Load	Demand factor	Diversity of groups
Transformer No. 1	10 kW	0.65	1.5
Transformer No. 2	12 kW	0.6	3.5
Transformer No. 3	15 kW	0.7	1.5

If the diversity factor among the transformers is $1 \cdot 3$, let us find the maximum load on the feeder.

Solution:

Figure shows a feeder supplying three distribution transformers.

Sum of max. demands of customers on Transformer 1,

$\qquad$ = connected load $\times$ demand factor $= 10 \times 0 \cdot 65 = 6 \cdot 5$ kW

The diversity factor among consumers connected to transformer No. 1 is $1 \cdot 5$,

$\therefore$ Maximum demand on Transformer 1 $= 6 \cdot 5 / 1 \cdot 5 = 4 \cdot 33$ kW

Maximum demand on Transformer $2 = 12 \times 0.6/3.5 = 2.057$ kW

Maximum demand on Transformer $3 = 15 \times 0.7/1.5 = 7$ kW

As the diversity factor among transformers is 1.3,

Maximum demand on feeder $= \dfrac{4.33 + 2.057 + 7}{1.3} = 10.3$ kW

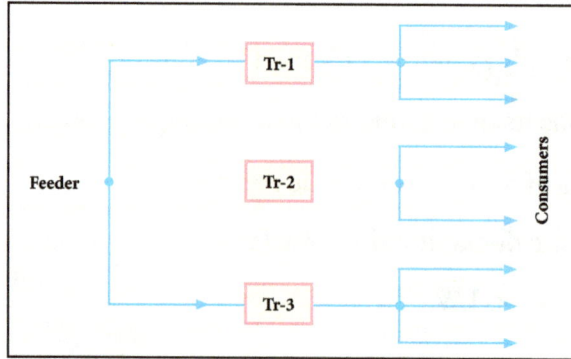

5. It has been desired to install a diesel power station to supply power in a suburban area having the following particulars:

- 1000 houses with average connected load of 1.5 kW in each house. The demand factor and diversity factor being 0.4 and 2.5 respectively.

- 10 factories having overall maximum demand of 90 kW.

- 7 tube wells of 7 kW each and operating together in the morning.

The diversity factor among above three types of consumers is 1.2. Let us know what should be the minimum capacity of power station?

Solution:

Sum of max. Demands of houses $= (1.5 \times 0.4) \times 1000 = 600$ kW

Max. demand for domestic load $= 600 / 2.5 = 240$ kW

Max. demand for factories $= 90$ kW

Max. demand for tube wells $= 7^* \times 7 = 49$ kW

The sum of maximum demands of three types of loads is $= 240 + 90 + 49 = 379$ kW. Since the diversity factor among the three types of loads is 1. 2,

$\therefore$ Max. demand on the station $= 379/1.2 = 316$ kW

$\therefore$ Minimum capacity of the station required $= 316$ kW

6. A generating station has the daily load cycle as follows,

Time (Hours)	0—6	6—10	10—12	12—16	16—20	20—24
Load (MW)	40	50	60	50	70	40

Now let us draw the load curve and find (i) maximum demand (ii) units generated per day (iii) average load and (iv) load factor.

Solution:

Daily curve is drawn by taking load along Y -axis and the time along X-axis. For the given load cycle, the load curve is shown in the following figure.

i. It is clear from the load curve that maximum demand on the power station is 70 MW and occurs during the period 16—20 hours.

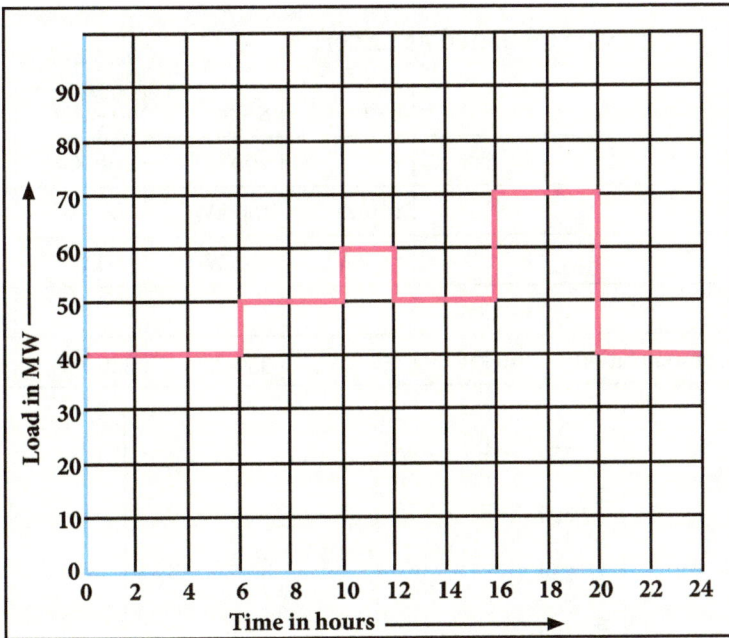

∴ The Maximum demand = 70 MW

ii. Units generated/day = Area (in kWh) under load curve,

$$= 10^3 \left[40 \times 6 + 50 \times 4 + 60 \times 2 + 50 \times 4 + 70 \times 4 + 40 \times 4 \right]$$

$$= 10^3 \left[240 + 200 + 120 + 200 + 280 + 160 \right] \text{ kWh}$$

$$= 12 \times 10^5 \text{ kWh}$$

iii. Average load $= \dfrac{\text{Units generated / day}}{24 \text{ hours}} = \dfrac{12 \times 10^5}{24} = 50,000 \text{ kW}$

iv. Load factor $=\dfrac{\text{Average load}}{\text{Max.demand}}=\dfrac{50,000}{70\times10^3}=0\cdot174=71.4\ \%$

7. A power station is in need to meet the following demand,

  Group A: 200 kW between 8 A.M. and 6 P.M.

  Group B: 100 kW between 6 A.M. and 10 A.M.

  Group C: 50 kW between 6 A.M. and 10 A.M.

Group D: 100 kW between 10 A.M. and 6 P.M. and then between 6 P.M. and 6 A.M. Let us plot the daily load curve and determine (i) the diversity factor (ii) the units generated per day and (iii) the load factor.

Solution:

The given load cycle may be tabulated as under:

Time (Hours)	0 - 6	6 - 8	8 - 10	10 - 18	18 - 24
Group A	-	-	200 kW	200 Kw	-
Group B	-	100 Kw	100 kW	100 kW	-
Group C	-	50 kW	50 kW	50 kW	-
Group D	100 kW	-	-	-	100 kW
Total load of power station	100 kW	150 kW	350 kW	350 kW	100 kW

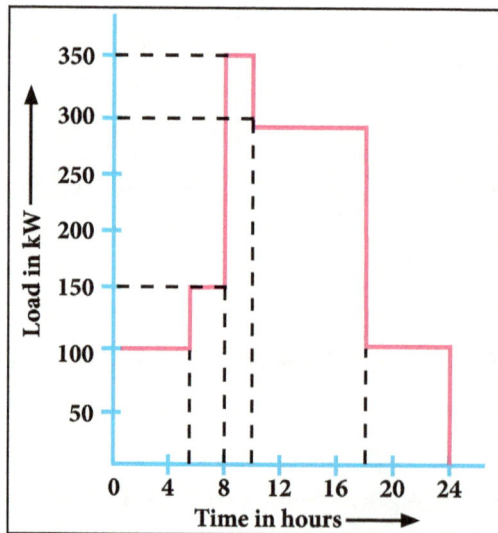

From this table, it is clear that total load on power station is 100kW for 0—6 hours, 150 kW for 6—8 hours, 350 kW for 8—10 hours, 300 kW for 10—18 hours and 100 kW for 18—24 hours. Plotting the load on the power station versus time, the daily load curve as shown in the figure. It is clear from curve that the maximum demand on the station is 350 kW and occurs from 8 A.M. to 10 A. M. i.e.,

Maximum demand = 350 kW

Sum of individual maximum demands of groups,

$= 200 + 100 + 50 + 100$

$= 450$ kW

Diversity factor $= \dfrac{\text{Sum of indivisual max. demands}}{\text{Max. demands on station}} = 450/350 = 1.286$

Units generated/day = Area (in kWh) under load curve

$= 100 \times 6 + 150 \times 2 + 350 \times 2 + 300 \times 8 + 100 \times 6$

$= 4600$ kWh

Average load $= 4600/24 = 191 \cdot 7$ k W

$\therefore$ Load factor $= \dfrac{191 \cdot 7}{350} \times 100 = 54 \cdot 8$ %.

Permissions

Index

www.ingramcontent.com/pod-product-compliance
Lightning Source LLC
Chambersburg PA
CBHW062008190326
41458CB00009B/3006